David Page

Geology for General Readers

A Series of Popular Sketches in Geology and Palaeontology

David Page

Geology for General Readers
A Series of Popular Sketches in Geology and Palaeontology

ISBN/EAN: 9783744675116

Printed in Europe, USA, Canada, Australia, Japan

Cover: Foto ©berggeist007 / pixelio.de

More available books at **www.hansebooks.com**

GEOLOGY

FOR

GENERAL READERS.

A SERIES OF POPULAR SKETCHES

IN

GEOLOGY AND PALÆONTOLOGY

BY

DAVID PAGE, F.R.S.E. F.G.S.

AUTHOR OF 'TEXT-BOOKS OF GEOLOGY AND PHYSICAL GEOGRAPHY,' 'HANDBOOK OF
GEOLOGICAL TERMS,' 'PAST AND PRESENT LIFE OF THE GLOBE,'
'PHILOSOPHY OF GEOLOGY,' ETC.

SECOND AND ENLARGED EDITION

WILLIAM BLACKWOOD AND SONS
EDINBURGH AND LONDON
MDCCCLXVI

PREFACE.

DURING the last twenty years immense progress has been made in the dissemination of Geological knowledge, yet still the old cry, "It is so difficult, so full of technicalities, so hard to be understood." Now, without admitting that Geology is fuller of technicalities, or one whit more difficult to be understood, than other sciences, I have thought it worth while, in the present volume, to attempt a simple and familiar exposition of its leading truths and principles. Because some have made the same attempt and failed is no reason why I should not try; because it is fashionable in some quarters to sneer at popular sketches is no reason why I should be deterred from expressing my conviction that sketches of this kind are the only means by which the majority of people can acquire any knowledge of science, while in many instances they form the first steps even to those who subsequently profess to despise them. Because information is given in a popular way, it need not be inaccurate and flimsy;

because it is told in a familiar manner, it need not be either undignified or unattractive.

Discarding technicalities as much as possible, and avoiding the formality of a text-book, I have still arranged the subjects so as to present a connected view of the science; and he who reads them in order, and understands as he reads, will have a fair conception of the scope and bearings of Geology. At the same time, each sketch is complete in itself, and contains, as far as it goes, an outline of our present knowledge of the matter to which it refers. This mode of treatment may lead to an occasional repetition, but a repetition in such cases is rather an advantage, as tending to render the matter more intelligible, and fix it more enduringly on the memory. I may also mention that several of the topics have been repeatedly brought before miscellaneous audiences in the form of popular lectures, and naturally indulge the hope that what was appreciated by them will prove equally inviting and instructive to the miscellaneous reader.

D. P.

GILMORE PLACE, EDINBURGH,
February 1866.

SECOND EDITION.

THIS Edition, so speedily called for, contains few new facts or discoveries, but it embodies three additional chapters — one on *Metamorphism*, or that internal change which every rock in the earth's crust is incessantly undergoing; another on *Veins*, their nature and origin; and a third on *What we owe to our Coal-fields*. Besides these chapters, which seem necessary in a compendium of General Geology, a number of emendations have been made throughout the work— the author's desire being to render it still more acceptable to that extensive and ever-increasing section of readers for whom it is specially intended.

October 1866.

CONTENTS.

I. THE CRUST WE DWELL UPON.

II. WASTE AND RECONSTRUCTION.

III. VULCANISM—ITS NATURE AND FUNCTION.

IV. METAMORPHISM, OR THE TRANSFORMATIONS OF ROCK-MATTER.

V. THE PRIMARY PERIODS.

VI. VEINS—THEIR NATURE AND ORIGIN.

VII. FOSSILS—THEIR NATURE AND ARRANGEMENT.

VIII. THE OLD RED SANDSTONE.

IX. COAL AND COAL-FORMATIONS.

X. THE OLD COAL-MEASURES.

XI. WHAT WE OWE TO OUR COAL-FIELDS.

XV. THE GLACIAL OR ICE EPOCH.

XVI. RECENT FORMATIONS.

XVII. MAN'S PLACE IN THE GEOLOGICAL RECORD.

XVIII. ORDER AND SUCCESSION OF LIFE.

XIX. WHAT OF THE FUTURE?

BRITISH STRATIFIED SYSTEMS,

TO WHICH THE FOLLOWING SKETCHES MORE ESPECIALLY REFER.

The subjoined tabulation exhibits the arrangement of the British stratified rocks, as accepted by our leading geologists—minor and local deviations of superposition being subordinated for the sake of distinct comprehension and ready reference :—

Systems.	Groups.	Periods.		
QUATERNARY.	In progress. Recent.			
TERTIARY.	Pleistocene. Pliocene. Miocene. Eocene.	CAINOZOIC.		NEOZOIC CYCLE.
CRETACEOUS.	Chalk. Greensand.			
OOLITIC.	Wealden. Oolite. Lias.	MESOZOIC.		
TRIASSIC.	Saliferous Marls. Muschelkalk (?) Upper New Red Sandstone.			
PERMIAN.	Magnesian Limestone. Lower New Red Sandstone.			
CARBONIFEROUS.	Coal-Measures. Millstone Grit. Mountain Limestone. Lower Coal-measures.			
OLD RED SAND-STONE.	Yellow Sandstones. Devonian Limestones and Slates. Red Sandstones, Conglomerates, and Cornstones. Grey fissile Sandstones ("flagstones") and Conglomerates.	PALÆOZOIC.		PALÆOZOIC CYCLE.
SILURIAN.	Upper Flags, Shales, and Limestones. Lower Grits and Slates.			
CAMBRIAN.	Slates, Schists, and Grits.			
LAURENTIAN.	Gneissic Schists, Quartzites, and Serpentines.	EOZOIC.		

OF VOLCANIC

RANGE OF TRAPPEAN ROCKS

RANGE OF GRANITIC ROCKS

ORDER AND SUCCESSION OF LIFE,

ESPECIALLY AS REFERRED TO IN THE FOLLOWING SKETCHES.

The subjoined tabulation exhibits proximately the stages at which the great ascending sections of Plants and Animals make their first appearance in the stratified systems:—

CAINOZOIC.	QUATERNARY.	Man. Plants and animals of existing species and distribution; a few genera recently extinct.	
	TERTIARY.	Placental mammals. Plants and animals of existing orders; a large proportion, however, of extinct genera and species.	
MESOZOIC.	CRETACEOUS.	Marsupial mammals, birds, reptiles, fishes, shell-fish, crustacea, zoophytes; palms, coniferæ, ferns, lycopods, sea-weeds.	
	OOLITIC. TRIASSIC.	Marsupial mammals, birds, reptiles, fishes, shell-fish, crustacea, zoophytes; palms, cycads, coniferæ, ferns, lycopods, sea-weeds.	
PALÆOZOIC.	PERMIAN. CARBONIFEROUS.	Reptiles, fishes, shell-fish, crustacea, zoophytes; coniferæ, ferns, lycopods, sea-weeds.	
	OLD RED SANDSTONE.	Fishes, shell-fish, crustacea, zoophytes; ferns, lycopods, sea-weeds.	
	SILURIAN.	Shell-fish, crustacea, worm-tracks, zoophytes; sea-weeds.	
EOZOIC.	CAMBRIAN.	Crustacea, worm-burrows, and zoophytes.	
	LAURENTIAN.	Traces of lowly or foraminiferal organisms.	

Vertical range labels (right side): RANGE OF INVERTEBRATA. RANGE OF VERTEBRATA. RANGE OF AMPHIGENS. RANGE OF ACROGENS. RANGE OF GYMNOGENS. RANGE OF ENDOGENS. RANGE OF EXOGENS.

GEOLOGY

FOR

GENERAL READERS.

THE CRUST WE DWELL UPON.

NATURE OF THE EARTH'S CRUST OR SOLID EXTERIOR—DIFFERS FROM
THE INTERIOR—COMPOSED OF ROCKS AND ROCK-FORMATIONS—
THESE THE THEMES OF GEOLOGICAL INVESTIGATION—TECHNICAL
MEANING OF THE TERM "ROCK"—STRATIFIED AND UNSTRATIFIED
ROCKS—HOW AND BY WHAT PROCESSES FORMED—OLDER AND
YOUNGER ROCKS—EXAMPLES OF—HOW DISTINGUISHED—CHRONO-
LOGICAL ARRANGEMENT OF ROCK-FORMATIONS—EACH A CHAPTER
OF WORLD-HISTORY—ATTRACTIVE NATURE OF THIS HISTORY—
FACTS ARRIVED AT BY A STUDY OF THE EARTH'S CRUST—THEIR
THEORETICAL AND PRACTICAL IMPORTANCE.

WHEREVER we travel we find the land made up of rocks
and rocky substances. If we go to the sea-shores, we find
similar substances stretching away beneath the waters, or
rising up in mid-ocean as reefs and islands. The fair inference therefore is, that all the exterior of our planet is
composed of rocks and rock-formations, and that the ocean
merely occupies the great hollows or depressions in the
same way as the lesser lakes and tarns occupy the rock-

B

basins of the continents and islands. To this rocky exterior geologists apply the term "crust," much in the same way as the housewife speaks of the crust of her loaf, or the schoolboy of the crust of ice that forms on the stagnant pool during the frosts of winter. The crust is something hard and consistent, and may differ both in nature and consistency from the interior on which it rests, or over which it may be formed; and this is precisely the idea entertained by geologists when they speak of the outer shell or "crust of the globe."

The rocky exterior over which we travel, and into which we dig and mine and tunnel, is a thing we can see and investigate to a limited depth; but the interior, sinking away four thousand miles to the centre, is placed altogether beyond our reach and observation. It may consist of rocky substances, but if so, they must be in a condition as to density altogether different from those we find at the surface; for as a planet the earth has a certain astronomically-ascertained weight, and were the force of gravitation to exert itself to the centre on such rocks as we know, their compression would give to the earth's mass a weight far exceeding that which its astronomical relations will allow.* Again, as we descend into the earth by mines, shafts, and Artesian wells, the temperature seems to increase at a given ratio (about one degree Fahr. for every 60 feet of descent); and at this rate a depth would soon be reached at which every known substance would be held in a state of incandescent fusion, or even vaporiform dispersion. It is convenient, therefore, to draw a distinction between the "crust" we can examine, and the "interior," respecting which we

* The reader must guard against the idea that at extreme depths all substances suffer alike from mutual mechanical pressure. Their different compositions forbid this supposition; and their densities must continue to depend (no matter what the depth) more on their chemical nature than on the amount of compression to which they are subjected.

can only form hypotheses.   And it is this crust which constitutes the great theme of geological investigation.   What is the nature of the rocks of which it is composed?   how are they arranged?   by what agencies have they been formed? what changes are they now undergoing?   and, reasoning from the known to the unknown, what changes do they seem to have undergone in former periods?   If we can answer all these questions, or approach to anything like a reasonable answer, we then present something like a history of our planet; and such a history is the aim and object of all sound geology.

We have said that this earth-crust consists of rocks and rock-formations; and here we must explain that the term "rock" is applied by geologists to all the solid substances that enter into its composition.   And there is good reason for this usage.   The sand and gravel of the sea-shore are but comminuted rock-matters derived from the cliffs above; the sands and loams and clays of the valleys are merely rock-debris, worn and washed in course of ages from the hills and uplands.   Be it gravel or sand, clay or mud, all are alike known to the geologist as "rocks;" and there can be no doubt that, were these loose and soft matters consolidated by pressure or other agency, they would become again compact and hard, like the rock-masses from which they were originally derived.   It is necessary, then, to bear in mind this technical use of the term "rock;" and the least reflection upon the changes (mechanical and chemical) which all rock-matter is incessantly undergoing will show the appropriateness of the application.

Understanding, then, what geologists mean by the term "rock," and bearing in mind that their labours are restricted to the accessible crust, let us inquire a little more narrowly into the nature of the rocks of which this crust is composed, and the modes of their arrangement.   Wher-

ever the structure of the crust is revealed—whether along
the cliffs of the sea-shore, in ravines worn out by rivers, in
shafts sunk for mining, or in railway-cuttings and tunnels
—we see the rock-masses arranged in two great ways. A
large and extensive class—like the sandstones and shales
and limestones—lie in layers or beds one above another; and
a second class—like the granites, greenstones, and basalts—
exhibit no lines of bedding or layers, but occur in vast and
indeterminate masses. These two modes of arrangement
may be seen in almost every railway-cutting and sea-cliff,
and must obviously have arisen from different causes. Now
the great maxim in geology is to reason from the known to
the unknown, and to appeal from the existing operations of
nature to the operations of the past. For, as was long ago
well remarked by Hutton, " when from a thing which is
well known we explain another which is less so, we then
investigate nature ; but when we imagine things without a
pattern or example in nature, instead of Natural History,
we write merely fable." Abiding by this method, we find
nature at the present day laying down, in every lake and
estuary and sea, layers of mud and sand and gravel, varying in thickness and continuity, according to the extent
of the areas and the magnitude of the in-flowing rivers ;
and were these layers consolidated, sand would form sandstone, gravel conglomerate, and mud shale. Here then,
as we cannot regard nature acting in time past otherwise
than at present, we are entitled to infer that all rocks in
the earth's crust occurring in layers have been formed
through and by the agency of water—that is, that they
are the sediments of former lakes and estuaries and seas,
the particles of which they are composed having been worn
down by water, transported by water, and deposited in
water. Hence all such rocks are regarded as *aqueous, sedimentary,* or *stratified,* and indicate that the areas they now

occupy were at some former period the sites of lakes, estuaries, and seas.

In a similar manner we seek to explain the origin and nature of the basalts and greenstones that rise up in homogeneous masses, and not in layers or bed above bed. And when we go to the volcano 'or burning-mountain, and observe the discharges of molten lava, which when cooled assume a structure scarcely distinguishable from that of the basalts and greenstones, we are equally entitled to infer that these have originated like lava, and consequently have been formed through and by the agency of fire; hence we regard them as *igneous* or *volcanic* if we refer to their origin, and *unstratified* if to their mode of arrangement. There are thus in the crust of the globe only two great categories of rocks—the aqueous or stratified, and the igneous or unstratified; the former produced through and by the agency of water, the latter through and by the agency of fire. The stratified, by the waterworn particles of which they are composed, and their sedimentary arrangement, layer above layer, give evidence of the forces that operate from without; the unstratified, by their crystalline texture and the manner in which they break through and derange the sedimentary strata, of the forces that exert themselves from within.

These two sets of rocks are being formed at the present day—the stratified or sedimentary in every lake, estuary, and sea, and the unstratified or eruptive around every active volcano. And, as nature's operations are incessant, such rocks must have been formed during all time—from the current hour back through untold ages. Did the watery forces—rains, rivers, waves, tides, and ocean-currents—alone prevail, the dry land would in course of time be worn and wasted down to one uniform level, over which the ocean might roll in uninterrupted continuity. But just

as certainly as waste and degradation are going on from
without, so the fiery forces—the volcano and earthquake—
are as incessantly operating from *within,* upheaving new
lands and mountains, and conferring on the whole new irregularity and diversity of surface.   The earth's crust is thus
held in equilibrium between these two opposing forces, *fire*
and *water* — between waste and degradation on the one
hand, and reconstruction and upheaval on the other.   In
this way former lands have been wasted and worn down,
and former estuaries and seas filled with the sediments ;
old continents and islands submerged beneath the waters,
and the sea-bed upheaved into newer lands.   The rocks of
the earth's crust are the only memorials of these repeated
changes ; and if geology is earth-history, it must endeavour
not only to decipher the changes they record, but to arrange
them in chronological sequence and connection.

When we look, then, at the changes now taking place
on the earth's crust, and the new rocks that are in process
of formation, we behold in them the exact counterparts
of what must have taken place during all former periods.
Winds, frosts, rains—springs, streams, rivers—waves, tides,
and ocean-currents—are ever weathering and wasting the
rock-matter of the globe; and the matter worn down is
borne by rivers to lakes and estuaries and seas, and there
deposited in layers of mud and clay and sand and 'gravel,
or further reassorted by the tides and currents of the
ocean.   Coral-reefs, shell-beds, and other masses of animal
origin, are also accumulating in various parts of the ocean ;
while peat-bogs, swamps, and forest-'growths are adding
analogous masses of vegetable origin to the land.   Hot-
springs and mineral-springs are also carrying matters in
solution from the earth's interior, and depositing these along
their courses ; while volcanoes are ever throwing out from
the same interior showers of dust and ashes, and masses of

molten lava. The earthquake also, which is but another manifestation of volcanic agency, is ever breaking up the rocky crust—here raising the sea-bed into dry land, and there submerging the dry land beneath the ocean—here rending and fissuring, and there producing inequalities and varieties of surface. Whatever is worn and wasted from one portion of the crust is laid down in another; there is nothing lost; but the interchanges and variations are interminable. The crust we dwell upon—stable and enduring as we are accustomed to regard it—is thus a thing of incessant change, protean in its superficial aspects, and ever-shifting in its terraqueous arrangements.

If the earth's crust be thus continually worn away in one district and reconstructed in another, some portions—like the lavas of Etna and the delta of the Ganges—must be comparatively recent, and others—like the Grampian Mountains and the coal-fields of Britain—of vast antiquity. In the former instances, the lavas and mud-islands are forming beneath our observation; in the latter, the formative processes have ceased, and no perceptible change has occurred for ages. To arrange the rock-formations of the earth into chronological order is one of the first duties of geology, for without this sequence there could be no history, and a connected history of the changes this crust has undergone is the great object of all geological investigation. By a rock-formation is meant the strata that have been deposited in any lake, estuary, or sea-area. The layers of mud, clay, marl, sand, and gravel which have filled up any ancient lake constitute a *lacustrine formation;* the sediments that are similarly deposited in estuaries an *estuarine formation;* and those deposited in seas, and subsequently upraised into dry land, a *marine formation.* In course of time, by pressure, chemical and other means, sands become *sandstones,* gravels *conglomerates,* clayey muds *shales,* calcareous muds *limestones,* muds largely impregnated with iron *ironstones,* and vegetable masses *coals ;* and it is in this way that the sediments of former lakes and estuaries and seas have become the rocky strata that now constitute the crust of the globe. As might be expected, there will often be every degree of admixture among these rocky strata, just as there is every degree of admixture and impurity among the sediments of existing seas and estuaries. There will be sandstones argillaceous, and sandstones calcareous ; shales bituminous, shales calcareous, and shales ferruginous ; limestones argillaceous, and limestones siliceous ; coals so pure as to burn away without leaving scarcely a trace of ashes, and others so stony as to be altogether unfit for fuel. The solid crust is, indeed, mainly made up of *mixed rocks*—that is, of arenaceous (sandy), argillaceous (clayey), calcareous (limy), siliceous (flinty), bituminous (coaly), ferruginous (iron-impregnated), and other similar compounds ; but whether these rocks be sedimentary sandstones, grits, conglomerates, shales, limestones, and ironstones, or fire-formed lavas, greenstones, basalts, and granites, the great object of geology is to distinguish between the older and newer, to arrange them in chronological order, and so arrive, if possible, at a knowledge of the geographical conditions which accompanied their formation. Nor is this endeavour in the least chimerical or uncertain ; for as was well remarked by Humboldt, now nearly half a century ago— "The superposition and relative age of rocks are facts susceptible of being established immediately, like the structure of the organs of a vegetable, like the proportions of elements in chemical analysis, or like the elevation of a mountain above the level of the sea. True Geognosy makes known the outer crust of the globe, such as it exists at the present day. It is a science as capable of certainty as any of the physical descriptive sciences can be."

As a general rule, the oldest formations will be the deepest-seated, or, in other words, the strata that lie beneath must be older than those that lie above them. Generally speaking, too, the older rocks will be harder and more crystalline in texture than the younger. There may be isolated exceptions to this, but the great fact holds good, that the older formations are really the more crystalline, and that this characteristic becomes less and less marked till we arrive at the recent and superficial layers of clay, sand, and gravel. Again, all the stratified rocks are less or more fossiliferous—that is, contain the petrified remains of plants and animals—and these *fossils*, as they are called, lead to pretty correct inferences not only as to the relative ages of formations, but as to the conditions under which they were deposited. And they do it in this way. Every lake, or estuary, or sea, imbeds in its sediments the remains of plants and animals that have either been drifted from the land by rivers or have lived and grown in the waters of deposit. As these remains get imbedded in the sands, clays, and calcareous muds, and excluded from the action of the air, they gradually undergo a change, become impregnated with mineral solutions, and in course of time are petrified, or converted into stony matter like the strata that contain them. At the present day the plants and animals entombed in the delta of the Mississippi differ widely from those entombed in the delta of the Ganges; and were these deltas subsequently converted into rock-formations, those plants and animals would afford evidence of the kind of life and climate that prevailed in their respective areas. It is in this manner that the fossils found in the earth's crust bespeak the conditions under which they lived—aquatic or terrestrial, fresh-water or marine, inhabitants of a cold climate or inhabitants of a genial one. They further afford the best of all evidence as to the rela-

tive ages of formations—the more recent containing the
remains of plants and animals nearly akin to those still
peopling the earth, while the more ancient contain plants
and animals that differ widely from the existing—and this
difference increasing with the age of the formation.

ᵀ Here, then, by means, *first*, of superposition, *second*, by
mineral composition, and, *thirdly*, by fossil remains, the
geologist can arrive at the relative ages of the rock-formations that constitute the earth's crust, and can arrange
them into sections and systems and periods, just as the
historian arranges the reigns and dynasties and periods of
human history. As the one speaks of ancient, medieval,
and modern times, so the other speaks of primary, secondary, and tertiary systems. As the one groups the populations of the world into ancient, medieval, and modern, so
the other groups its life into eozoic, palæozoic, mesozoic, and
cainozoic—that is, dawn-life, ancient-life, middle-life, and
recent-life. But while the geologist thus reads the history
of the earth mainly through its stratified rocks, he at the same
time receives important aid from its igneous or unstratified
masses. These, as volcanic products, break through the
stratified formations, throw them out of their horizontal
position, overflow them in part, insert themselves among
them as intrusive masses, and fill up rents and fissures in
the form of dykes and veins. All this gives ample evidence
of former change, and presents a lively picture of the operation of these gigantic forces which are still so instrumental
in modifying the existing aspects of our planet. In fine,
the whole crust is replete with evidence, physical and vital,
of the earth's former conditions. Our globe writes, as it
were, her own history—every layer of mud and sand laid
down in water, every shower of ashes or sheet of lava
ejected from a volcano, every stem and twig, every shell
and tooth and scale preserved in sediment, forming an

incident in that progress which it is the great object of geology to unfold. Geology is in fact the Physical Geography of former ages. For just as the geographer endeavours to depict the existing aspects of sea and land, the climates they enjoy, and the plants and animals by which they are peopled, so the geologist labours to recall the aspects of the past—the distributions of sea and land at each successive stage, the plants and animals by which they were characterised, and by inference the nature of the physical conditions, genial or ungenial, by which they were surrounded. The methods of the one are but the methods of the other; and the more the geologist knows of the existing operations of nature, the better will he be able to interpret the operations of the past. The phenomena of the present are patent, and for the most part explicable; those of the past are obscure, and, in proportion to their distance and obscurity, the greater the interest excited and the ingenuity required for their interpretation.

And surely if men take an interest in the history of their own race—in the mounds and barrows, the tombs and pyramids, the towers and temples of bygone populations, whose dates extend at most to a few thousand years—much more ought to be their enthusiasm in that higher history which carries the inquirer from the historic to the prehistoric, and beyond the prehistoric into events and aspects whose distance can only be indefinitely indicated by eras and cycles. The events of the one history are scattered over a small portion of the earth's surface, and for the most part only under a few feet of rubbish; the events of the other are universal, and found in every stratum that enters into the composition of the rocky crust. The events of the one history are no doubt more direct and immediate; but the remoteness of the other, their strangeness and their variety, should only excite our interest the more, and exalt our conceptions of that Creative Wisdom which has exerted itself in this world of ours ages before the human race became witnesses of its beauties or participators in its bounties. And the clearer we can render this history, the more minute our analysis of the earth's crust; and the closer the connection we can establish between the successive stages of its formation, the more attractive and instructive will the geological record become. ⟋

In arranging the rock-formations of the earth, we may either divide them, as the older geologists did, into *Primary, Secondary,* and *Tertiary ;* or looking, as modern geologists do, more especially at their fossils, we may adopt the subdivisions, *Eozoic, Palæozic, Mesozoic,* and *Cainozoic.* In either case these main divisions contain several formations of marine, estuary, or lacustral origin, and these it is customary to name either after their prevalent rocks, their most characteristic fossils, or some geographical area in which they are typically displayed. Thus the Cretaceous or Chalk system is so named from chalk-rock forming its most distinctive feature in the south of England, and the Old Red Sandstone from its consisting largely of reddish-coloured sandstones; while the Silurian is named after the district between England and Wales, where it is typically developed, and which was anciently inhabited by the Silures, and the Laurentian because typically displayed in the region of the St Lawrence. Adopting this plan (and it matters little what the nomenclature, provided we be certain of the chronological order), the stratified rocks of the crust may be arranged in the following manner—not going into minutiæ, but simply presenting such leading features as may convey to the miscellaneous reader some idea of the sequence that prevails among the stratified systems, and the ascent of life, vegetable and animal, as it makes its appearance from the lowest to the highest formations :—

CHRONOLOGICAL SCHEME OF THE EARTH'S CRUST.

Era	Formation	Rocks	Life
CAINOZOIC.	QUATERNARY.	Sands, gravels, lake-silts, peat-mosses; coral-reefs, and other recent superficial accumulations.	Man. Plants and animals still existing; a few genera recently extinct.
	TERTIARY.	Stratified clays, sands, marls, various limestones, beds of lignite, &c.	Plants and animals of existing orders; a large proportion, however, of extinct genera and species.
MESOZOIC.	CRETACEOUS.	Chalk with and without flints, calcareous clays, lignites, greensands, and various sandstones.	Marsupial mammals, birds, reptiles, fishes, shell-fish, crustacea, zoophytes; palms, coniferæ, ferns, lycopods, sea-weeds.
	OOLITIC.	Calcareous sandstones, sandstones, clays, shales, coals, ironstones, and limestones.	Marsupial mammals, birds, reptiles, fishes, shell-fish, crustacea, zoophytes; palms, cycads, coniferæ, ferns, lycopods, sea-weeds.
	TRIASSIC.	Reddish and variegated sandstones, marls, pebbly beds, and shelly limestones.	
PALÆOZOIC.	PERMIAN.	Reddish sandstones and conglomerates, magnesian limestones.	Reptiles, fishes, shell-fish, crustacea, zoophytes; coniferæ, ferns, lycopods, sea-weeds.
	CARBONIFEROUS.	Sandstones, shales, fire-clays, coals, limestones, and ironstones, in repeated alternations.	
	OLD RED SANDSTONE.	Various coloured sandstones (often red), conglomerates, marly shales, and concretionary limestones.	Fishes, shell-fish, crustacea, zoophytes; ferns, lycopods, sea-weeds.
	SILURIAN.	Slaty beds, limestones, shales, grits, and sandstones.	Shell-fish, crustacea, worm-tracks, zoophytes; sea-weeds.
EOZOIC.	CAMBRIAN.	Slaty beds, crystalline schists, grits, and conglomerates.	Crustacea, worm-burrows, and zoophytes.
	LAURENTIAN.	Crystalline schists, limestones, and serpentines.	Traces of lowly or foraminiferal organisms.

Studying the preceding scheme, it will be seen how numerous are the formations that compose the earth's crust, each formation representing the sediments of former lakes and seas, and each varying in composition according to the conditions under which it was deposited. It will further be seen that from the oldest to the most recent there has been an ascent, in general terms, from the lower to the higher forms of life—the sea-weed preceding the fern, the fern the conifer, the conifer the palm, and the palm the true exogenous timber-tree; and so in like manner the zoophyte preceding the shell-fish, the shell-fish the fish, the fish the reptile, the reptile the bird, and the bird the mammal. We have thus revealed by a study of the earth's crust, what our forefathers never dreamt of—namely, that this crust is in a state of incessant change, what was formerly dry land becoming the sea-bed, and what was once the sea-bottom being upraised into dry land; that these old sea-sediments constitute the formations which compose the earth's crust; that these formations are replete with the evidences of former life; that this life evinces a progress from lower to higher forms; and that all the interchanges of sea and land, all the waste and reconstruction, all the growth and decay of bygone life, establish an antiquity for this world of ours vast beyond all human conception.

Summing up, then, our knowledge of the rocky crust—and this without any conjecture as to the nature of the earth's interior—it may be stated in general terms, *first*, That this rocky crust is in a state of slow but ceaseless change, and that the causes—meteoric, aqueous, igneous, chemical, and organic — that now waste and reconstruct have been productive of similar changes in all time past. *Second*, That these changes, like all other natural operations, must be governed by imperative laws, and that the mineral structure of the globe arising therefrom has consequently a definite and determinable arrangement. *Third*, That this arrangement, as displayed in the numerous rock-formations, implies an enormous lapse of time—time to waste and wear, time to transport, and time to deposit and reconstruct—and therefore establishes an antiquity for our globe vast beyond all previous conception. *Fourth*, That during the long periods which these successive formations—that is, successive distributions of sea and land—imply, the earth has been peopled by different races of plants and animals—all evidently belonging to the same great scheme of life, but varying widely in their characteristics during each succeeding epoch. *Fifth*, That during these periods there has been an ascent, in the main, from lower to higher forms ; and that the plants and animals now inhabiting the globe are, on the whole, higher and more specially organised than the plants and animals of any former period. *Sixth*, That these successive appearances and distributions of plants and animals are connected together in one great scheme of life by some pervading law of development which, though not yet satisfactorily discovered, is evidently bound up with the operating forces of the universe. And, *lastly*, The earth being still subjected to the same causes of change and, from all we can see, to the same law of development that operated in time past, the future aspects of our planet must differ from the present physically and vitally—its present distribution of sea and land giving place to other arrangements of sea and land, and its present living races to others of a still higher and more specialised organisation.

Such is the crust we dwell upon, and the teachings which a study of its structure can convey. This rocky exterior is all we know with certainty of the composition of our planet—the foundation of all geographical diversity, the diversified habitat of plants and animals, the scene of man's own life-labours, and the storehouse of those minerals and metals upon which his civilisation and progress are so intimately dependent. The study of its structure is replete with intellectual interest of the most exalted description. The variety of its rock-formations, the minerals and metals they contain, their modes of aggregation, and the curious changes to which they have been subjected in their repeated alternations from sea to land and from land to sea, are all calculated to excite our interest and increase our admiration of the means employed by the Creator to alter, to diversify, and to sustain. And that interest and admiration are increased a hundredfold when we perceive in these formations the nature of the life that has preceded us rising through long ages from the simple to the more complex, from the simply sentient to the intellectual and reflective, and this through forms so countless and varied, and yet all belonging to the same great plan, that numerous as are the existing forms of plants and animals, they form but a tithe of those that have necessarily existed before them. No one, then, can look into the structure of this crust without receiving newer and deeper insight into the laws and ordainings of nature, and from all deeper insight of nature the human intellect arises wiser, happier, and more exalted.

But the study of the earth's crust is not less desirable from its economic advantages than from its intellectual interest. Man's civilisation and progress, and his mastery over the powers of nature, are intimately dependent upon his knowledge and application of the minerals and metals. Indeed, modern civilisation and progress have largely arisen from this knowledge of the minerals and metals; and as these hold determinate positions in the various formations of the crust, an acquaintance with that crust is indispensable to their acquisition. Gold and silver, coal and iron, gems and precious stones, are not scattered indiscriminately through the earth. Some occur more abundantly in one formation than in another; some in beds, others in veins; some exclusively in one kind of matrix, others in another; and all this knowledge as to abundance, depth of strata, direction of veins, and the like, can only be acquired by a study of the structure and arrangement of the rocky crust. Geology has thus all the interest of a wondrous Past to attract; it possesses all the value of a sterling Present to incite to its study and acquirement. To the general reader its revelations of world-history will ever form themes of intelligent attraction; to the miner, the engineer, the architect, and others whose business is to deal with the structure and products of the rocky exterior, its deductions are of direct and special importance.

Such, once more, are the economic and intellectual advantages arising from a study of the structure of the crust we dwell upon—economic advantages of which our country, in every department of its industry, is every day reaping the benefit, and intellectual promptings which have led to a newer and deeper insight into the laws and ordainings of nature. We say newer and deeper insight, for with increased knowledge of the past must extend our knowledge of the present; and the tendency of all true knowledge of . God's workings in nature must ever be to make men better, wiser, and happier in all their relations to that nature of which they form so prominent a part. Everything is bound up one with another in the Divine scheme of the universe; and he who perceives this truth most fully in the physical world is surely the most likely to regard it in the intellectual and moral. On this ground alone, and altogether independent of its intellectual pleasures and economic advantages, the science of this earth-crust is worthy of our closest cultivation—leading the mind from the harmonies that prevail in the natural world up to the higher harmonies that ought to pervade the human and social.

WASTE AND RECONSTRUCTION.

" OUR solid earth is everywhere wasted, where exposed to
the day. The summits of the mountains are necessarily
degraded. The solid and weighty materials of those mountains are everywhere urged through the valleys by the force
of running water. The soil, which is produced in the destruction of the solid earth, is gradually travelled by the
moving water, but is constantly supplying vegetation with
its necessary aid. This travelled soil is at last deposited
upon the coast, where it forms most fertile countries. But
the billows of the ocean agitate the loose materials upon the
shore, and wear away the coast, with the endless repetitions
of this act of power, or this imparted force. Thus the continent of our earth, sapped to its foundation, is carried
away into the deep and sunk again at the bottom of the
sea, whence it had originated, and from which, sooner or
later, it will again make its appearance. We are thus led
to see a circulation in the matter of this globe, and a system

of beautiful economy in the works of nature." Such are
the words of Dr Hutton in his celebrated 'Theory of the
Earth,' towards the end of last century; and such the
conclusion at which every one must arrive who gives the
matter sufficient and enlightened consideration. But this
incessant transmutation of the solid framework of the globe
is a conception not readily realised by ordinary minds, partly
from the restricted range of observation during a single life-
time, and partly from our limited notions of time, which is
in itself illimitable and altogether independent of the events
that mark the course of its continuity. This difficulty was
not unforeseen by the Scotch philosopher, and so he goes
on to remark :—" It is not to *common* observation that it
belongs to see the effects of time, and the operation of
physical causes, in what is to be perceived upon the surface
of the earth. The shepherd thinks the mountain on which
he feeds his flock to have always been there, or since the
beginning of things ; the inhabitant of the valley cultivates
the soil as his father had done, and thinks that this soil is
coeval with the valley or the mountain. But the man of
scientific observation, who looks into the chain of physical
events connected with the present state of things, sees great
changes that have been made, and foresees a different state
that must follow in time, from the continued operation of
that which actually is in nature." It is the object of the
present Sketch to place this system of waste and reconstruction—of destruction and renovation—in a clear and obvious
light, that the " common " as well as " scientific" mind may
perceive the means employed by the Creator to keep this world
of ours ever young notwithstanding its vast antiquity, and
to maintain its stability in the midst of incessant vicissitude.

To the casual observer the hills and valleys that surround
him appear unchanged and unchangeable. The plains and

battle-fields mentioned in ancient history, the sites of cities and harbours, the courses of rivers, and the contour of mountains, are much the same as when described one thousand, two thousand, or even four thousand years ago.   But to him who looks a little more narrowly the case is altogether different.   The stream in the valley has cut for itself a deeper channel, and has repeatedly shifted its course—eating away the banks on one side, and laying down spits of new ground on the other.   The cliffs in the hills are more weather-worn and rounded, and a larger mound of rock-debris has accumulated at their bases.   The lakes of the old historic plain are partly converted into marshes, and the marshes into meadow-land ; the site of the old city on the sea-cliff has been partly wasted away by the encroaching waves ; and the ancient harbour, once at the river-mouth, is now a goodly mile inland, and separated from the sea by a flat alluvial delta.   The Nilotic plain is not precisely the same as when described by Herodotus ; the sunderbunds or mud-islands of the Ganges have been largely augmented during the last two hundred years ; and many areas that were laid down on the charts of our earlier traders as mud-flats, now form fertile portions of the great Chinese plain.   Vesuvius has repeatedly changed its aspects since Herculaneum and Pompeii were buried beneath its ejections ; and there is scarcely an active volcano that has not materially added to its bulk since the commencement of the current century.

Such changes are incessant, and though individually they may seem insignificant, yet when viewed in the aggregate, and continued from century to century, they assume a magnitude commensurate with the crust of the globe itself, every portion of which has repeatedly suffered degradation and renovation, been repeatedly spread beneath the waters as sediment, and as repeatedly reconstructed into newer strata and upheaved into dry land.   Imperceptibly as the rains and frosts may wear away the mountain-cliff, slowly as the river may deepen its channel, gradually as the delta may advance upon the estuary, and little by little as the volcano may pile up its scoriæ and lava, yet after the lapse of ages the mountain will be worn down, the river-channel will be eroded into a valley, the estuary converted into an alluvial plain, and the volcano rear its cold and silent dome into the higher atmosphere. All that is necessary is time, and this is an element to which we can see no limit in the future, any more than we can discover a beginning to it in the past. To render these incessant mutations thoroughly intelligible, however, to the ordinary observer, it will be necessary to describe the agents by which they are effected, and at the same time the varying power of these agents according to the latitudes and altitudes within which they operate. These agencies may be conveniently arranged under five great categories; namely, 1, The *Meteoric*, or those—like winds, rains, and frosts—depending upon the atmosphere; 2, The *Aqueous*, or those—like rivers, waves, and tides—arising from the action of water; 3, The *Chemical*, or those resulting from chemical actions and reactions; 4, The *Organic*, or those—like peat-mosses and coral-reefs—depending on the growth and decay of plants and animals; and 5, The *Igneous*, or those—like the volcano and earthquake—connected with the manifestations of heat within the interior of our planet. Each of these agencies has its own mode of working—some chiefly wearing and degrading, some degrading and at the same time accumulating, and others solely reconstructing. Let us now glance at them in detail :—

The principal effect of the *Meteoric* or *Atmospheric* agencies is to weather and wear away. Slowly but surely the gases and moisture of the atmosphere eat into every exposed rock-surface. The disintegrated matter is washed down by the rains, taken up by the runnels and streams, and borne onward by the rivers to the ocean. We often see the effect of heavy rainfalls on exposed soils and surfaces in our own islands—how they batter, loosen, and carry away; but our rainfall, amounting annually to some 30 or 40 inches, is trifling compared with the rainfall of tropical and sub-tropical countries, ranging from 200 to 400 inches, and this concentrated for the most part within one period of the year. It is not uncommon to hear travellers speak of the soils being converted into mud, and of the rivers running mud rather than water, and this solely through the battering and dissolving influence of the periodical rains. Again, frost in all the higher latitudes and altitudes is annually performing a similar function. The moisture that inserts itself into the pores and interstices of all rock-substances is converted into ice during frost; ice occupies more space than the water of which it consists, or, in other words, water expands during freezing; the particles of rock-matter are distended or forced asunder; and when thaw comes, their cohesion being loosened, they are washed away by the rains and carried down by the streams and rivers. Every winter we see the disintegrating effects of frost on the ploughed soils, road-cuttings, and sea-cliffs of our own islands; and this effect is manifested a hundredfold in all the colder latitudes and in all the higher mountains, whether within tropical, temperate, or arctic regions. The destructive power of frost is stupendous, whether silently crumbling away the cliffs and precipices; discharging the avalanche and landslip down the mountain-slope; slowly grinding its way as the glacier through the Alpine glen; or transporting and dropping, as the iceberg does, its burden of rock-debris over the floor of the ocean. As with the rains and frosts, so to a certain extent with the winds or aërial currents of the atmosphere. Wherever there is rock-matter sufficiently light and loose, thence the

winds will remove it and carry it away to some more sheltered locality.   And if the set of the wind be constant, or
chiefly from one direction, like the trades and sea-breezes,
the result in the long-run will be very marked and perceptible.   By this means the dry sand of the sea-shore is
blown inland and beyond the reach of the tide into mounds
and hillocks (sand-dunes, as they are termed), and along
every shore in the world there are recently-formed expanses
of this nature, often—like the "Landes" of France—of vast
extent, and still in the process of augmentation.   As with
the sands of the sea-shore, so with the sands of the arid
deserts; they are driven hither and thither into dunes and
ridges, but chiefly forward in one main direction according
to the prevailing winds, and this to the obliteration of
streams and oases, and to the destruction of fertile valleys
that lie in their way.   Gentle as it may seem, the drifting
of sand over the surface of granite and basalt has been
known to wear and polish down their asperities, and even
to grind out grooves and furrows like those produced by
the long-continued motion of glacier-ice or the flow of running water.*

But perceptible as may be the effects of the meteoric
forces, they are far less obvious than those produced through
and by the agency of water.   The *Aqueous* are generally on
a larger scale ; and wherever streams and rivers run, waves
break and tides ebb and flow, there they are to be witnessed,
partly as degrading, but partly also as accumulating and reconstructing forces.   The mere passage of water over rock-
surfaces would of itself have little effect ; but as it bears

---

* At the Pass of San Bernardino in California, Mr W. P. Blake (as
quoted by Professor Dana) observed the granite rocks not only worn
smooth, but covered with scratches and furrows by the sands that were
drifted over them.   Even quartz was polished, and garnets were left
projecting from pedicles of felspar.   Limestone was so much worn as to
look as if the surface had been removed by solution.

along its burden of sand and gravel and shingle, every particle becomes a tool which grinds, and is in turn ground down in the double process of attrition and erosion. Every runnel and rivulet wears for itself a channel, and bears the eroded material down to the river; the river performs the same operation, but on a larger scale, and with marked intensity during floods and freshets, cutting out ravines and gorges, or scooping out broader valleys, and transporting the debris to the lower levels of lakes, estuaries, and the ocean. There the mud and sand and gravel borne from the higher grounds come at last to rest, subside as sediments, and are thus spread out as alternating strata, to be consolidated by pressure, chemical agency, and other means, and ready, when the event happens, to be upraised as the rock-formations of newer lands. In like manner, also, with the waves and tides and currents of the ocean. Restlessly and for ever eating into and undermining the sea-cliff, the waves encroach upon the land, pound down the hardest material to shingle and gravel and sand, and this with rapidity according to the nature of the opposing cliff, and the manner it is disposed to the impact of the breakers. The effects of wave-action are perceptible along every exposed shore; here in the undermined and falling cliffs, there in caverns and gorges, and in another part in the "needles" and outstanding rock-masses that have been severed from the land. What the waves have worn down the tidal ebb and flow disintegrate still more, and scour and carry away to the stiller depths and more sheltered recesses. And the great ocean-currents, too—like the Arctic with its burden of icebergs and rock-debris, or the Gulf Stream with its drifted sea-weeds and animal exuviæ—are also incessantly transporting and reassorting. Everything, however, comes at last to rest in the waters, being either piled on shore as sand, gravel, and shingle, deposited as silt in the deeper and stiller waters, or strewn along the courses of the ocean-currents in long reaches of miscellaneous debris, partly of animal, partly of vegetable, and partly of mineral origin. The effect of aqueous agency is thus partly to wear and waste, and partly to accumulate and reconstruct—to wear down the old continents and to accumulate the abraded materials in the waters for the formation of newer lands.

The *Chemical* agencies, though less perceptible, are not less general or less incessant in their action than the meteoric or aqueous. Indeed, in a certain sense the meteoric and aqueous act chemically ; but we have hitherto alluded chiefly to their mechanical effects, and now direct attention more especially to their chemical. The carbonic acid of the atmosphere eats into the most crystalline marble ; its oxygen converts the hardest ironstone into a soft earthy peroxide, ready to be washed away by the first shower that falls. Every spring that issues from the interior of the earth holds in solution, less or more, some mineral or metallic matter, which it either deposits along its course or carries forward through stream and river to the ocean. Be it lime, or iron, or flint, or salt—and such matters constitute our petrifying, chalybeate, siliceous, and saline springs— these matters must have been dissolved or wasted from the interior, as they are now brought to its surface to form new rock-masses or to enter into newer combinations in the waters of the ocean. And this chemical effect of springs is vastly increased when the waters are hot, whether bursting forth like the Geysers of Iceland, or simmering like the mud and sulphur vents that appear in the neighbourhood of almost every active volcano. Heat, indeed, is the great promoter of chemical change within the earth's crust ; and from this cause arise, no doubt, those discharges of naphtha, petroleum, and the like, that result from the slow decomposition of lignites, coals, and other organic masses. It is not to pressure alone, nor to volcanic heat alone, that the solid strata, originally of sand, gravel, mud, and organic debris, owe their hardness and crystalline texture. Chemical infiltrations and combinations are everywhere as operative as these are, and, indeed, in most instances are the main modifiers of mineral texture, colour, and consistency. And the veins and veinstones—the great repositories of the metallic ores—that traverse the older formations, they, too, are the immediate products of chemical segregation, slowly and silently, but ever at work in these secret recesses. On the whole, chemical actions and reactions within the rocky crust of the earth are incessant, either dissolving and displacing, reconstructing into other forms, or aggregating in other and newer compounds.

The *Organic* agents fall next to be considered, and of these, as of many other departments of nature, it may be remarked that the minute and unobserved are the most active and effective. It is true that the trees of the forest may be imbedded in peat-bogs or drifted into the mud of estuaries, and that the bones of fishes, reptiles, birds, and mammals may be entombed in the sediments of lakes and seas; but, on the whole, these constitute a small proportion of the containing strata, and even where swept by currents into special shoals and bone-beds, as we know they are in certain parts of the ocean, they form but insignificant accumulations compared with those resulting from the myriad-growths of microscopic organisms. In the vegetable world the plants of the peat-moss and swamp-growth claim our first attention, as out of the constituents of the air and water they elaborate their own substance, and year after year bequeath it to the accumulating mass. How thin soever may be the film that the growth and decay of a single year may add to the accumulation, yet in the course of centuries the peat-growth and swamp-growth thicken, till now within all the colder

latitudes there are expanses of vast area and many feet in
depth entirely composed of decayed vegetation, and ready
to be converted into coal, like the analogous accumulations
of former epochs.   In the animal world, on the other hand,
the coral-reef is perhaps the most striking instance of aggregation by the minutest of means.   Barely perceptible to
the naked eye, the tiny zoophyte, in countless myriads,
secretes the lime held in solution by the waters of the ocean,
till year after year and century after century the conjoint
structure so increases that at last the " reef " stretches away
many leagues in length and many fathoms in thickness.
As with coral-reefs so with beds of gregarious and drifted
shells, and so also with the enveloping limy and flinty
shields of microscopic organisms like the foraminifera and
diatoms ; the former mere specks of animal jelly, the latter
mere points of plant-life, yet so increased by the immensity
of their numbers and the rapidity of their growth, that large
areas of the sea-bed as well as lakes and estuaries are thickly
strewn with their calcareous and siliceous envelopes.   These
envelopes, though mere microscopic specks, are yet aggregated
in such myriads that they are capable of forming extensive
beds of limestone on the one hand, and of flinty rock on
the other.   And such limestones and flint-rocks we find in
the solid crust, exhibiting to the eye of the microscopist
the beautiful organisms of which they are composed, and
proving that in former ages the earth's crust was built up
by the same agencies that still continue to remodel and up-
hold it.   A great proportion of the chalk-rocks of England,
the nummulitic limestone that stretches from the Alps eastward through Europe and Asia on to the Philippine Islands,
and the mineral known as tripoli or polishing-slate, are
ancient strata formed by analogous microscopic organisms ;
and the same work still goes forward in the calcareous ooze
that covers so large an area of the Atlantic sea-bed, and in

the siliceous mud of most of our existing estuaries.* The lime and the flint dissolved by springs from the crust of the earth, and borne down by streams and rivers to the ocean, is thus secreted by vital agency and once more converted into solid rock-matter. Nothing is lost; it may change its shape and pass out of sight for a while, but in the long-run it will reappear, altered it may be in form, but essentially the same in substance. Every particle of matter obeys a ceaseless round of change. Now in the crystalline and independent gem, at another time as a constituent of the solid rock; now dissolved in the limpid waters, at another time built up in the structures of plants and animals; now scattered abroad as the decaying exuviæ of life, and once more collected into compact and rocky strata.

The *Igneous* agents generally exert themselves with signal force and marked effects, and yet in some instances their most gigantic results are brought about by stages that are almost imperceptible. The most notable instance of their operation is perhaps in the volcano, which in course of ages piles up its alternate discharges of dust and ashes and lava till it assumes the lofty proportions of an isolated mountain like Etna, or stretches away in long ranges like the fiery cones of the Andes. Whether exerting itself on land or rising up from the depths of the ocean, the volcano is one of the most important modifiers of the earth's crust; and it acts partly by upheaval, partly by accumulation, and partly by fusing and reconstructing rock-matter in the interior, and by bringing it once more to the surface. In this function it

---

* Of the microscopic organisms—the foraminifers, polycystines, diatoms, and desmids—that stand, as it were, on the confines of Life, the two former belong to the animal kingdom and the two latter to the vegetable. The foraminifera secrete calcareous matter, the polycystines and diatoms siliceous, and the desmids no appreciable quantity of either. The foraminifera and diatoms, along with the coral-polypes, may therefore be regarded as the main microscopic rock-builders.

is usually accompanied by the earthquake, which fractures and dislocates the solid crust, uplifting one portion and depressing another, submerging one area beneath the waters and elevating another into dry land. Unlike the volcano, the earthquake produces no direct change on the character of the rocks; but it is the great diversifier of the earth's surface, creating new irregularities, and as a consequence interfering with the action of the atmospheric and aqueous forces that operate on that surface, as well as with the distribution of the life by which it is peopled. But beyond the earthquake and volcano, with all their minor accompaniments of hot-springs, mud-springs, and the like, there seems to be another and still more gigantic, though silent, manifestation of vulcanic power. We allude to those gradual uprises and depressions of portions of the earth's surface—those crust-motions, if we may so speak—by which certain regions, like the shores of Scandinavia, Spitzbergen, and Siberia, are being slowly raised above the sea-level, and other regions, like the western coast of Greenland and that of the Southern States of America, as slowly depressed beneath it. Many countries, and our own islands among the rest, are marked by lines of ancient sea-beach, denoting uprise above the waters; and could we only see beneath the ocean, we believe there are other regions equally marked by terraces of depression. There is no other power save Vulcanism, or internal heat-force, to which we can ascribe these upheavals and depressions of the crust; but be this as it may, there can be no doubt of such oscillations, and that they form one of the appointed means by which the waste and reconstruction of our continents are held in equilibrium, and new distributions of sea and land accomplished.

The effect of these various agencies—the atmospheric, aqueous, organic, chemical, and igneous—is thus not only to mould and modify the exterior, but at the same time to build up the interior structure of our planet. Their function is at once to wear down the old and to reconstruct the new; to scatter abroad in one region and to accumulate in another. And the new rocks reconstructed must necessarily bear the closest relationship to the old from which they were derived. Their hardness and compactness and crystalline texture is merely a matter of time. Given time and the fitting conditions, and the loosest sand will be converted into the most compact sandstone, the softest mud into the hardest slate, and the earthiest chalk into the most crystalline marble. Though here separated for the purposes of elucidation, these agents are ever working hand in hand—the atmospheric with the aqueous, the aqueous with the chemical, and the chemical with the organic and igneous. And it is this complicated working that renders the composition of the solid crust so varied, the aspect of its rocks so different, and the task of unravelling their history at once the trial and the triumph of geology.

Such is a brief sketch—a mere indication as it were—of the great forces by which the earth's crust is incessantly modified, its rock-matter wasted and reconstructed, and the equilibrium of its terraqueous distribution sustained. Sketchy as the outline has been, the careful reader will have perceived not only the nature of the modifying agents, but the manner in which they operate; and must feel convinced that, small as may be their results over a given area or during a given time, yet comprising the whole globe and allowing for ages, they are sufficient to accomplish any amount of change—to destroy, in fact, the whole of the existing continents and to reconstruct new ones from the bed of the ocean, and this by degrees, and over and over again, according to the course of Time, which is illimitable and beyond all computation. He will also have perceived

that, many as are the agents at work and complicated as are
their modes of action, yet, on the whole, they may be conveniently arranged into two grand categories—namely, the
powers of waste and degradation from without, and the
powers of reconstruction and upheaval from within. As
surely as the meteoric and aqueous disintegrate and level,
so surely does the igneous reconstruct and upheave ; as
the chemical and vital dissolve and destroy in one area, so
they recombine and build up in another. There is nothing
so harmonious as this incessant round of mutation, nothing
so marvellous as the variety it produces, and yet nothing
so certain as the unity of design by which the whole is
combined into one intelligible system. Strange as it may
seem, even the comfort and development of man is indissolubly bound up with this system of vicissitude. It is
not the rugged and flinty hill-side that yields him his sus-
tenance. He cannot build his cities on its peaks or plough
its precipices. And yet these hills are the great store-
houses of future fertility. The rains, the frosts, the streams,
and the rivers are perpetually carrying down from their
heights the materials to form the fertile valleys—washing
out from the old crystalline rocks the inorganic elements
indispensable to vegetable luxuriance. And on these moun-
tain-derived plains man has hitherto settled in communi-
ties and built his cities. The plains of the Old World,
the historic fulfilments of the past—China, Hindoostan,
Mesopotamia and Egypt—were borne down from the
mountains of Asia and Africa ; just as the prairies, the
llanos, and pampas of the New World, the hopes of the
advancing future, are the gifts of the Andes and the Rocky
Cordilleras. How marvellous this system of interdependence between the organic and inorganic — between the
mechanical processes of nature and the social development
of man ! How admirable this system of unceasing rejuven-

escence ! the old hills worn and wrinkled and furrowed by decay, and the younger valleys spreading out in their beauty and freshness and fertility !

" This earth, like the body of an animal," said Hutton, " is wasted at the same time that it is repaired. It has a state of growth and augmentation; it has another state, which is that of diminution and decay. This world is thus destroyed in one part, but it is renewed in another; and the operations by which this world is thus constantly renewed are as evident to the scientific eye as are those in which it is necessarily destroyed." And yet how few will take the trouble to comprehend this system of incessant change, contented to live in the belief that this earth has always been as it is, has been so from the beginning, and will continue to be so to the end ! Verily they deprive themselves of much rational enjoyment, pay little regard to the system of nature of which they form so prominent a part, and show little reverence for Him who has given them eyes to see, and understanding to understand, so be it they will only learn to exert them. How admirable the system of compensation by which decay in one part is balanced by renovation in another! The substances disintegrated by water are again reconstructed by fire ; the matter dissolved by chemical action is collected anew by vital; and what was appropriated for a while by the living organisms is restored again to the mineral world when vitality has ceased its requirements. Everything in this universe is indissolubly woven into a network of interdependence, and not a mesh could be taken away without destroying the beauty and consistency of the whole.

> " From nature's chain whatever link you strike,
> Tenth, or ten-thousandth, breaks the chain alike ! "

And the reason is obvious; for most of the operating forces we have described arise directly from the earth's primal connections as a member of the solar system. Annual revolution and daily rotation, daily and seasonal alternations of heat and cold, currents and counter-currents of wind, evaporation and rainfall, waves and tides, currents and counter-currents of water, are all results of the earth's relation to the sun and her sister-planets. So in like manner are the phenomena of vegetable and animal growth; so, too, are many of those chemical and electro-magnetic activities with which science is but slenderly acquainted; and so also, perhaps, though in a remoter degree, that manifestation of internal vulcanism concerning which philosophy can do little more than merely hazard conjectures. So long, therefore, as the earth's primal relations endure, these secondary forces must operate as a necessary consequence, and thus the rocky crust must continue to undergo a round of waste and reconstruction as ceaseless as the revolutions of the planetary system, and as permanent in its power. How harmonious the system by which this earth, in the midst of all its mutations, is kept ever fresh and young! Day after day, and year after year, the aspects are ceaselessly changing, but the vitality remains the same; cycle after cycle the forces may shift their direction, but their power remains unimpaired. Everything around us being seemingly stable, it may be difficult to realise this conception of incessant change, just as it is impossible to estimate the lapse of time required for its fulfilment; but an effort must be made, and not till the mind has learned to form some idea of the ceaseless mutations to which the earth's crust is subjected, the causes by which these mutations are effected, and the amount of time required for their production, can it be said to comprehend the fundamental truths upon which the science of geology is erected.

VULCANISM—ITS NATURE AND FUNCTION.

WHATEVER be the nature and origin of the thermal forces
that operate within the crust of our earth—whether deep-
seated or near the surface—whether arising from chemical
actions or dependent on some primordial condition—it is
convenient to arrange them under one general term, and
that of *Vulcanism* or *Vulcanicity*, suggested by Humboldt
in his ' Cosmos,' seems by far the most comprehensive
and appropriate. In this way, not only the volcano proper, but the earthquake, hot-springs, gas-springs, mud-
springs, and all kindred phenomena, are brought under one
category; and "it is really advantageous," as remarked by
the great German philosopher, "to avoid the separation of
that which is causally connected, and differs only in the
strength of the manifestation of force, and the complication
of physical processes." Here, then, we give to the constantly active reaction of the interior of the earth upon its
external crust or surface the name of *Vulcanism;* and the

object of the present Sketch is to describe the operations of
this internal heat, to explain what is known of its origin,
and to define its apparent function in the economy of nature.   To render a subject intelligible it is necessary to be
methodical ; and we shall therefore, without any assumption
of technical exactitude, arrange all the thermal phenomena
of our globe under the three great heads of *Volcanoes,
Earthquakes,* and *Crust-Motions.*

Every one is less or more acquainted with the aspect and
nature of a *Volcano* or burning-mountain.   Whether he has
seen one or not, his readings and hearings lead him to associate with his idea a mountain of a conical form, having a
crater or orifice of eruption at top, and from which at intervals are emitted clouds of vapour and flame, showers of
dust and ashes, and streams of *lava* or molten rock-matter.
The outline may not be strictly conical, yet such is the
form which it usually assumes ; there may be more craters
or orifices than one, some being *central* or near the top, and
others *lateral* or placed along the sides ; and the substances discharged may be very heterogeneous—now highly
heated steam and sulphurous vapours, now dust and cindery
matters called *scoriæ,* now fragments of rock (lapilli and
volcanic bombs), and anon wellings-out of lava, sometimes
extremely fluid and at others slaggy and cindery.   Notwithstanding such local and periodic differences, there is, on the
whole, a great similarity in aspect and operation between
all volcanoes, which leads to the belief that they belong to
one brotherhood, and to the same system of causation.   It
is usual to speak of *lava-cones, tufa-cones, cinder-cones,* and
mixed cones, according as a mountain is chiefly composed of
one or other of these substances, or of a mixture of all of
them; but, as might be anticipated, the mixed cones are by
far the most prevalent, and the distinction is mainly valu-

able in assisting the observer to determine the composition of distant or inaccessible volcanoes. Thus the slope of a lava-cone is very gentle—from 3 to 10 degrees; that of a tufa-cone from 15 to 30 degrees; of a cinder-cone from 35 to 45 degrees; and that of a mixed cone usually gentle beneath, but topped with a steep peak of loose and scoriaceous materials. Again, some are strictly *sub-aërial*—that is, take place on the dry land—and others *sub-aqueous*—that is, operate under the waters, or but rarely manifest themselves at the surface; yet both seem to act in a similar way, and to discharge similar products. Further, some are ceaselessly *active;* others become active only at long intervals, and are said to be *dormant;* while others have been so long dormant and shown no symptoms of activity that they are regarded as *extinct.* Between the existing and the extinct there is every grade of activity, just as among the extinct there is every degree of antiquity. We are thus led from the active craters of Etna and Vesuvius back to the extinct cones of Central France and the Rhine, which, in their crateriform domes and rugged lava-streams, still retain the aspect of volcanoes; from these back through the Apennines and Alps; from the Alps to the Pyrenees; and from the Pyrenees to the Scandinavian and older mountain-ranges, which have all had the same origin; though now their craters and domes are obliterated, and their outlines have undergone a thousand modifications from those denuding agencies of air and water which have operated upon them during untold ages. And here the reader cannot be too strongly impressed with the fact that *the profiles of all our existing hill-ranges are more the results of waste and denudation from without, than of upheaval and accumulation from within.* When a mountain is first presented to us, the natural idea is, no doubt, that of upheaval and accumulation, but a little reflection will soon correct the misconception. All the older lands have been repeatedly under the sea, and have suffered at each depression and re-elevation the denuding effects of wave and tidal action. Even since their latest re-elevation, the rains, frosts, and rivers have been incessantly wearing, wasting, and eroding —so much so, that the older mountain-ranges are but the merest skeletons of what they once were. Vulcanic agency may block out, as it were, the contours and profiles of the land ; but the meteoric and aqueous agents—the frosts, rains, rivers, and waves—are the busy chisellers that are for ever conferring upon it its latest features—the *latest*, but never the last.

In looking upon the more ancient hills, as well as upon existing volcanoes, a question naturally arises—Are these elevations chiefly upliftings of the earth's solid crust, or are they accumulations of igneous matter that have been dis-charged from its interior ? In other words, are mountains and mountain-ranges mainly produced by upheavals or swellings-up of the earth's rocky crust, or have they been accumulated on the surface by repeated discharges of vol-canic matter ? Much controversy has existed on this point, and many arguments adduced on both sides ; but the truth seems to be, that the forces from within have acted in both ways—partly by elevation of the stratified crust, but chiefly by the accumulation of erupted materials. In this view every mountain and mountain-range becomes a matter of slow and gradual growth, every shower of ashes and stream of lava adding to the bulk of the isolated cone, and every new cone adding another link to the mountain-chain. We have no exact measure of this slow and gradual accumula-tion, but judging from the small amount that has been added to Etna and Vesuvius during the historical period, many of the existing volcanoes must be of vast antiquity ; and when we carry our retrospect back through the extinct volcanic hills to the ancient mountain-ranges, the mind is altogether unable to grasp the cycles that must have elapsed since their formation.

Presuming, then, that volcanic hills are chiefly masses of accumulation and not of upheaval, the repeated eruptions that take place must necessarily fracture and derange the continuity of the surrounding district, and thus every igneous centre is marked by such accompaniments as hot-springs, boiling mud-springs, discharges of sulphurous gases, and the like, better known perhaps as the *suffioni* and *solfataras* of Italy, and the *salses*, *hornitos*, and *hervideros* of Mexico and South America. Such minor discharges are the normal accompaniments of all active volcanoes, and long after activity has ceased they form the residual phenomena, and indicate by their declining force and numbers the distance both in time and place of the fiery forces that once operated below. No doubt springs of considerable temperature may exist in districts long since quiescent, and now far removed from volcanic activity (those of Bath and the Pyrenees, for example); but the monticules thrown up by mud-volcanoes and escapes of heated and sulphurous vapours generally mark either the proximity of igneous activity or the comparative recentness of its manifestations in the area. The whole are merely indications of the same thermal agency—that internal fire-force which Humboldt has so appropriately included under the name of Vulcanism or Vulcanicity.

The next great manifestation of vulcanism is the *Earthquake*, a distinction made in scientific as well as in every-day language; for though the earthquake is generally the close concomitant of the volcano, yet its throes may be felt in districts where no volcano has existed for ages. This motion, as the name implies, is a quaking or trembling of

the earth—varying from the gentlest tremor to agitations
so violent that the solid crust is fractured, one portion
thrown up and another thrown down, the sea-bed uplifted
into dry land, and the dry land submerged beneath the
waters.   The phenomena that accompany earthquake convulsions are extremely varied.   Occasionally they are preceded by an unusual stillness and sultriness of the atmosphere;
low hollow rumblings, more audible than felt; and great
restlessness and terror among birds and mammals, as if the
instincts of these were keener than human perception.   At
other times there is no premonition, but all at once a few
smart concussions, passing away in a certain direction, or
not unfrequently spreading from a central point in diminishing intensity.   On other occasions, however, the mo-
mentary concussions return after a short pause with increased
vehemence, and then there is a perceptible undulation of
the earth's crust, as if it were passing away from beneath
the feet of the spectator—an uplift, a shock, a series of
giddy shocks, and the work of destruction.   Though con-
tinuing at most for a few seconds, these violent shocks
generally result in extensive fracturing of the rocky crust.
Yawning rents and fissures, gaseous discharges, bursting
forth of new springs, absorption of streams, changing the
course of rivers, elevation of the sea-bed, submerging of
the dry land, and the conversion of populous cities into
masses of ruin and rubbish, are the not unfrequent but
destructive effects of the earthquake.   At times too, as the
water in a vessel that has been agitated and then brought
suddenly to rest strikes forcibly over its margin, so the sea,
its floor having been shaken, is frequently thrown into
violent waves (earthquake-waves), which rush forward
against the land to the height of 40, 50, or 60 feet, and
sweep everything into destruction before them.   The wave
that rolled in upon the coasts of Portugal after the great

Lisbon earthquake in 1755 was estimated at 60 feet high, and the succession of such waves (three in number) that desolated the town of Simoda (Japan) in 1854 were little inferior in violence and dimensions.

On the whole the effects of the earthquake are much more disastrous than those of the volcano. The discharges of the one, being at considerable altitude, are chiefly felt for a few miles round its crater or in long narrow streams down its sides; but the other convulses for leagues, and this at all levels and alike over land and over sea. The two, however, are usually in close connection; and in centres of igneous activity, when the volcano begins to discharge, the convulsions of the earthquake cease, or at all events lose much of their intensity. The one acts as a sort of safety-valve to the other, and this necessarily so if we regard them as both arising from the same deep-seated source of igneous intensity. This connection was long ago noticed by Dr Hutton, who quaintly but somewhat bitterly remarks—" A volcano is not made on purpose to frighten superstitious people into fits of piety and devotion, nor to overwhelm devoted cities with destruction. A volcano should be considered as a spiracle to the subterranean furnace, in order to prevent the unnecessary elevation of land and fatal effects of earthquakes. And we may rest assured that they, in general, wisely answer the end of their intention, without being in themselves an end, for which nature had exerted such amazing power and excellent contrivance."

The third great manifestation of the reaction of the earth's interior upon its external crust consists in those slow movements by which certain portions of the land are stage by stage elevated above the waters, and other portions as gradually depressed beneath them. To this manifestation we may apply the name of *Crust-Motion*, as

indicating a slow and long-continued movement of the
solid crust in the region where it occurs, in contradistinction to the volcano and earthquake, whose operations
are sudden and convulsive. Whether these gradual crust-
movements result from the same igneous forces that give
rise to the earthquake and volcano is a matter open to
question, but in the present state of our knowledge we can
perceive no other adequate cause, and are therefore compelled to associate them with the same pervading vulcanicity. Along the coasts of our own islands, and indeed
along the coasts of every other country, the attentive observer will perceive at various levels above the present
sea-beach several shelves or terraces, which meet the eye
like reaches of former shore-line. On closer inspection,
their parallelism, the sand and gravel of which they are
composed, the shells, bones, and other marine exuviæ which
they contain, prove incontestably that the sea formerly
stood at these levels, and that the land to this extent has
been successively elevated above the waters. Whether
such elevations took place suddenly or by slow degrees it
is often impossible to tell, but it is readily seen from the
old beaches and the cliffs that guard them that the sea
must have long stood at their successive levels. If the
movement takes place suddenly—like that which during
the present century elevated the coast of Chile to the
height of eight or ten feet, or that which depressed the
Run of Cutch, or that which more recently uplifted part of
the north island of New Zealand to the height of six feet or
thereby—it is generally ascribed to earthquake convulsion ;
but if it occurs by slow stages it is regarded as crust-motion,
the proximate cause of which is at present unknown.

As notable instances of this crust-motion we may point
to the shores of Scandinavia, which have been long known
to be rising at a slow and equable rate; to the raised beaches

of Spitzbergen, as described by Mr Lamont and other recent visitors; to the ancient shore-lines of Siberia, as amply illustrated by Von Wrangell; the numerous terraces of uprise noticed by Dr Kane and other Arctic voyagers on the coasts of Greenland and the Arctic islands; the terraced shores of Patagonia, long since observed by Mr Darwin; as well as to the less distinct, because more ancient, shore-lines that encircle our own islands and the opposite coasts of France and the Spanish Peninsula. As with these uprises, so also with several depressions that have been noticed, though these from their nature are generally less perceptible. Such tracts of subsidence have been observed along the west coast of Greenland, the southern coasts of the United States, and generally in the basin of the South Pacific. Whatever the nature of these uprises or depressions—whether at the rate of four or five feet a century, as in the case of Scandinavia, or with greater or less rapidity; whether recent, like those of Spitzbergen and Greenland, or ancient, like those of our own country—they all belong to the same class of phenomena, and are evidently the result of some great but unknown law. Physicists have attempted an explanation, some attributing the phenomena to oscillations of the hypothetical molten interior, and others to secular expansions and contractions of portions of the crust, as arising from changes in the axis of rotation and centre of gravity. In either case the oscillation of a few thousand feet is insignificant as compared with the diameter of the globe; and as elevation in one region seems to be counterbalanced by subsidence in another, the general relations of our planet may be regarded as standing for ages unaffected by the amount of its superficial changes.

Insignificant, however, as may be the effect of such oscillations upon the general relations of the earth, they are all-important to the climate, and consequently to the flora and fauna of the region in which they occur. A few hundred feet of elevation or depression in the higher latitudes of the north is tantamount to a loss of several degrees of annual temperature ; and we can readily conceive what would be the effect of another thousand feet of uprise on the existing flora and fauna of Siberia, Greenland, or the Arctic islands of America. In fine, it requires no great stretch of the imagination to conceive to what extent the climatology of the globe may be influenced by a system of extensive elevations and depressions of its surface ; how the cold and warm currents of the ocean might be diverted; how one region might be elevated so as to be permanently enveloped in snow and ice, while another in the same latitude might lie at so low a level as to enjoy the amenities of a temperate climate. On the whole, whatever may be the origin of these slow and gradual crust-motions, we behold in them a system by which the distribution of sea and land is changed, by which climate is modified, and consequently by which the plant-life and animal-life of our planet is materially affected.

Such are the three great manifestations of vulcanism— the volcano, the earthquake, and the gradual crust-motion; and though their origin be obscure, the human mind seldom rests satisfied with mere description, but must attempt a solution of cause and origin. There are two principal hypotheses that have been advanced to account for vulcanic phenomena, and which may be respectively termed the *mechanical* and the *chemical*. By the former, the whole is resolved into an aboriginal igneous condition of the earth's mass, on which, after the lapse of ages, a cooled and rocky crust has been formed over a molten interior. To oscillations in this molten interior, to its reactions upon the crust, to the cavernous structure of the crust, and to

chinks and fissures that admit the percolation of water
down to the incandescent mass, are ascribed the tremors
and convulsions of the earthquake and the sudden explo-
sions of the volcano. By the latter hypothesis it is presumed that the solid crust contains abundance of metallic
elements, such as potassium, sodium, calcium, magnesium,
and the like, and that the percolating waters coming in
contact with these produce instantaneous combinations,
which result in uncontrollable manifestations of heat, and
the conversion of these metals into their oxides—potash,
soda, lime, magnesia, &c.—which enter largely into the
composition of the rock-matter ejected at the surface. Such
is a brief and general view of the two leading hypotheses
that have been advanced to account for vulcanic phenomena. The adherents of the former question the presence
of these metallic elements in such abundance as to produce
such gigantic results, and point to the universality of volcanic action and the uniformity of its products as evidences
of its arising from the same great interior source. The
adherents of the latter contend for a system of action and
reaction, without which the globe would gradually lose its
supposed interior heat, and become so cool that in process
of time volcanic action would cease—a result incompatible
with the maintenance of a diversified and habitable surface.
According to the mechanical theory, say they, the interior
heat must be gradually declining, and must finally come to
an end; but according to the chemical, there is a round of
incessant action and reaction, a system of compensation
and endurance which accords with the other ordainings of
the universe. This is not the place to do more than merely
allude to these contending views, our object in these sketches
being rather to explain what is known than to discuss
what is questionable.

But whatever be the origin of vulcanic force, we see it

abundantly manifested in various regions of the globe—
here in isolated centres like those of the Mediterranean
and Iceland, there in linear directions like the Andes, and
occasionally over wide areas like the Indian and Chinese
Archipelagoes. At the present moment there are between
three and four hundred active volcanoes (with as many
more in a dormant or semi-extinct state), chiefly fringing,
as it were, the Pacific or scattered over its surface. In the
latter instance they appear in insular centres, as in the
Sandwich and other groups; in the former they occur in
great lines, as in the Andean, Mexican, and Columbian
Mountains, the Aleutian Islands, Kamtschatka, Japan, the
Philippine Islands, and the Indian Archipelago. Compared with these the display of volcanic energy in other
regions is insignificant; while over immense tracts like
the north of Asia and Europe, and the Atlantic slopes of
both Americas, the internal forces have been still and stationary for ages. We know from the mountain-ranges of
these regions that the forces once were there; we see their
effects in the primary granitic mountains, in the secondary
hills of basalt and greenstone, and in the tertiary domes of
trachytic lava;* but of the law that has regulated this shifting from area to area, and now restricted it to its present

---

* It may be of use to the general reader to mention that the principal
rocks in recent volcanic hills are—*lavas* of various aspect and compactness; *tufas*, or consolidated cindery matters; *pumice*, or light vesicular
lava; *obsidians*, of glassy texture; and *trachytes*, or granular-crystalline
masses : and that their differences depend partly on chemical composition, but chiefly on the rapidity with which they have been cooled—
rapid cooling producing a compact glassy texture, and slow cooling the
reverse. The rocks of the secondary hills, on the other hand, though
originally consisting of the same volcanic ejections, are now converted
into crystalline *greenstones, basalts, felstones,* the softer *trap-tuffs,* and
amygdaloids, or those whose original vesicles have got filled with almond-
shaped infiltrations of lime-spar, agate, and other minerals. In the older
mountains the conversions are still more complete, and highly crystalline
granites, syenites, and *porphyries* are the prevailing compounds.

centres, we know next to nothing, and are merely on the threshold of the inquiry.

Little, however, as we know either of the cause or of the course of vulcanism, we can readily perceive its function, and behold in it one of the great means by which the crust of the globe is held in equilibrium, and by which the diversity and variety of its surface is maintained. Were there no adequate force acting from within, the powers of waste and degradation from without would in time reduce the surface to one dreary monotony of level, incompatible with that diversity of condition and of life which appears to be one of the great aims of creation. But just as the meteoric and aqueous forces wear and waste from without, so the vulcanic renovate and upheave from within; and thus the rocky crust is held in perfect equipoise, and its surface diversified by all that irregularity of hill and valley, of table-land and plain, which is indispensable to variety in the plant-life and animal-life by which it is adorned. Wherever large expanses of the earth's surface, like the prairies and pampas of America, are characterised by sameness of condition, there is a consequent want. of variety in their vegetable and animal existences; but as the great design of Creation seems to be variety in space as well as variety in time, this uniformity of surface is incessantly broken up by the operations of the earthquake and volcano. Locally disastrous as may be the throes of the one or the discharges of the other, we thus behold in each a necessary part of the world's mechanism, and powerful in proportion to the work it has to accomplish.

It has been frequently discussed, and with some is still a question, whether this power of internal vulcanism be steadfast or declining, and whether it did not manifest itself with greater intensity during the earlier geological periods? Locally we may perceive that it has ceased in some areas, and in others seems gradually declining; but at the same time we behold it breaking forth in new areas, and on a survey of the whole world, see no reason to conclude that it is now either less extensive in its distribution or diminishing in its intensity.   The Andes, through whose extreme length the earthquake and volcano are ceaselessly active, are as gigantic as the Himalaya, where they have long since ceased to exist; the Mexican Cordilleras broader and loftier than the Alps; the Alps more imposing than the older Pyrenees; and the Pyrenees as decided in their character as the primitive ranges of Scandinavia.   The cincture of volcanic action that girdles the Pacific (in the Andes, Californian Sierras, Aleutian Isles, Kamtschatka, Japan, Philippine Islands, the Indian Archipelago, and New Zealand, to say nothing of the groups that stud its bosom) is as gigantic as any axis or area geology has revealed,* while the individual discharges and irruptions are unexcelled by those of any former period: and in corroboration of this we need only point to the lava streams of Etna, from twenty to forty miles in length; to those of Iceland, full fifty miles in length, by twelve to fifteen in width; or to

* In absence of a map the following arrangement may convey to the reader a general idea of the disposition of these volcanic *lines* and *centres*. —1. *A long the borders of the Pacific :*—Throughout the entire length of the Andes from Tierra del Fuego northwards; in Central America; Mexico; Oregon; the Aleutian Islands; Kamtschatkan peninsula; the Kuriles; Japan group; Philippines; East India Islands; New Guinea; East coast of Australia and New Zealand.   2. *Over the Pacific :*—In the Sandwich Islands; Friendly Islands; Fejees; Santa Cruz group; New Hebrides; and the Ladrones.   3. *Over the seas that lie between the northern and southern continents, and adjacent regions :*—West India Islands; the Mediterranean and its borders; the southern borders of the Caspian and eastward; and the East Indian Archipelago as lying between Asia and Australia.   4. *In the Indian Ocean :*—Bourbon and the Mauritius; Comoro group; and Madagascar.   5. *In the Atlantic :*—St Helena; the Cape Verdes; Canaries; Madeira; Azores; and Iceland.   The interiors of the great continental masses both in the old and new worlds are still and quiescent.

those of the Sandwich Islands, nearly seventy in length, and of varying width and thickness, according to the nature of the declivity down which they have flowed. On the whole, there seems no ground for the supposition that vulcanism is now either less extensive or less energetic than in former ages. All that geology perceives is that it has shifted from area to area; that the old mountains where it once reigned are long since cold and silent; that the secondary hills bear but few traces of its presence; and that now it upheaves along other lines and convulses from newer centres.

That in vulcanism, as in other cosmical forces, there must be a law determining both its time and mode of operation is sufficiently obvious; but in the mean time that law lies altogether beyond the indications of science, and we must content ourselves with mere descriptions of phenomena and explanations of function. What the cause of volcanic energy? what the periodicity of its discharges? what the times and directions of earthquake convulsions? and what the law which regulates the shifting of vulcanism from centres that were once disturbed to newer areas? are questions to which science can give no satisfactory answer, and generations may pass before even the way to a solution is indicated. What we want is, *first*, a more minute investigation of the products of volcanoes, that it may lead to a knowledge of the decompositions taking place within the earth's interior; *second*, observations as to the connection between earthquake throes and volcanic discharges; *third*, a record of the times and directions of earthquake convulsions; and, *fourthly*, an accurate series of tabulations and maps to assist us in our deductions. All this is little more than merely beginning to be done. Chemistry and physics have only recently been directed to the subject; and Seismology or Seismography (the science of earth-shocks) is a study as it were but of yesterday.* Till substantial progress has been made in this direction, and sufficient facts for generalisation collected, it were idle to speculate, and waste of time to surmise.

Such is a brief, and we trust not unintelligible sketch of that Vulcanicity or internal heat-force which is incessantly reacting upon the rocky crust we inhabit. Than the earthquake, volcano, and great crust-pulsation, we have no higher manifestations of natural force ; no phenomena before whose power man's weakness becomes more apparent. There are, no doubt, other terrific agencies in nature—the ocean when lashed into fury by storms, the flooded and headlong river, the hurricane, and the thunderstorm. Man, however, learns to brave and battle with these. The hardy islander dares the ocean-storm in his little skiff; civilised nations build their piers and breakwaters that their fleets and navies may ride behind them in defiance of the storm. Man dams and diverts the river-current, restrains it within bounds, or even turns it to account as the moving power of his machinery. By strength and weight of material he can resist the fiercest sweep of the wind-blast ; or if need be, can yoke it to his wheels submissive and serviceable. He even toys with the thunder, and brings the lightning down from the storm-cloud. But before the shock of the earthquake and the throes of the volcano, man—savage or civilised— shrinks altogether abject and helpless. With them, however frequently they may occur, he never becomes familiar. The earth, with which all his ideas of stability are associated, rocks and reels beneath him—his proudest cities become an instantaneous mass of ruin and rubbish ; himself falls prostrate, or if he flees, he flees only to accelerate his fate.

* See the ' Seismology,' and ' Seismographic Maps,' of the Messrs Mallet, of Dublin. 1861.

The volcano casts forth its scorching showers of scoriæ and ashes, his pastures and vineyards are utterly consumed, and his homesteads and villages—like Pompeii and Herculaneum —are buried, so that for centuries their very places are unknown. Or the red river of lava spreads slowly and irresistibly down the mountain-side, crushing and consuming the forest-growth like stubble, damming and diverting river-courses, engulfing villas and towns, and converting the fair face of nature into a wilderness of blistery slag and "the blackness of desolation."

Opposed to these terrific agencies, man stands utterly impotent; and alas for his ideas of this Scheme of Creation, if he cannot learn to associate them with its necessary order and mechanism! Exposed to incessant powers of waste and degradation from without, the earth's crust is only maintained in habitable equilibrium and variety of surface by this power of vulcanism acting as incessantly from within. As it now manifests itself by the three or four hundred orifices known to geographers, so in all time past have its manifestations been equally apparent—now in this region, now in that, now feebly, now with greater intensity, fitful to all appearance, but obeying, we may rest assured, some fixed and determinable law, could we only grasp the multifarious conditions associated with its expression. As at present, so during all the geological periods, we find that it has been gradually elaborating hills and mountain-ranges, upheaving and consolidating the sediments of the sea-bottom, giving diversity of surface to areas of flat and uniform deposit, raising substances of utility from inaccessible to accessible depths in the earth's crust, and reticulating that stony fabric with lines and walls of injected matter, which under the names of "dykes" and "faults" restrain the percolation of its internal waters, and bring them to the surface in thousands of living and refreshing streams. The great diversity of animal and vegetable life which now adorns this earth is clearly associated with its diversity of surface, and this diversity of surface—this arrangement into hill and dale, into table-land and plain, into island and continent, with all its varying accompaniments of wind and ocean-currents and climate—is the direct result of the earth's incessant vulcanicity. Unless then, when considering the earthquake, volcano, and crust-motion, we can learn to regard them as necessary portions of the orderly mechanism of our planet, we are viewing them in the feeblest light, and much as the cowering savage, who sees in them only disorder and destruction, and the wrath of his offended deities. And even where science cannot pierce the veil, we may rest assured, in this as in other arrangements of nature, that nothing is jarring, but that all, however unintelligible to our finite comprehensions, is in perfect harmony, and indispensable to the maintenance of this wondrous world-plan—

> " All seeming discord, things not understood ;
> All partial evil, universal good."

METAMORPHISM, OR THE TRANSFORMATIONS OF ROCK-MATTER.

METAMORPHISM, DEFINITION OF THE TERM—ALL ROCK-MATTER
INCESSANTLY UNDERGOING INTERNAL CHANGE THROUGH PRES-
SURE, ATTRACTION OF COHESION, CHEMICAL ACTION, HEAT, MAG-
NETISM, CRYSTALLISATION, AND OTHER SIMILAR FORCES —
ILLUSTRATIONS OF THESE RESPECTIVE FORCES AND THEIR MODES
OF ACTION—THEIR EFFECTS MOST PERCEPTIBLE IN THE OLDER
ROCKS—THE SO-CALLED METAMORPHIC ROCK-SYSTEM—APPARENT
OBLITERATION OF ITS FOSSILS—GRADUAL DETECTION OF THESE IN
PECULIAR LOCALITIES—RESOLUTION OF THE SYSTEM INTO MINOR
SECTIONS AND CHRONOLOGICAL ORDER—INCONCEIVABLE AMOUNT
OF TIME IMPLIED IN THE DEPOSITION AND SUBSEQUENT TRANS-
FORMATIONS OF THE METAMORPHIC SYSTEM—HOW TO DEAL
WITH IT AS A PART OF WORLD-HISTORY—ITS CONVENIENCE AS
A PROVISIONAL DESIGNATION.

IT has been already explained that all the substances composing the earth's crust, from the most recent and superficial soils, sands, muds, and gravels, down to the oldest and
deepest slates and schists, are known as rocks and rock-formations. It has also been shown that all these rock-matters
are undergoing incessant changes, being weathered and worn
from the old hills, borne onwards by streams and rivers,
laid down as sediments in lakes and seas and estuaries, and
again upheaved by vulcanic agency as the solid strata of
newer continents. In this long and ceaseless round they are
disintegrated and dissolved, reconstructed and consolidated,
and it is to the varied processes of reconstruction and consolidation that we would here direct attention. Every substance,

the moment it is laid down as sediment or discharged from
the crater of a volcano, begins to suffer change and transformation.   The attraction of cohesion, pressure, the percolation
of chemical solutions, heat, magnetic currents, crystallisation, and the like, are all more or less altering its internal
texture, and in the long-run its external aspect or structure.
In course of time, and under the operation of these forces,
the softest mud becomes compacted into shale, sand into
sandstone, gravel into conglomerate, peat-moss into coal,
and coral-growths into limestone ; and by a farther transformation shales may become glistening clay-slates, sandstones quartzites, coals anthracites, and limestones sparkling
and. crystalline marbles.   This kind of transformation, or
Metamorphism, as it is technically termed, forms one of the
most wonderful as it does one of the most difficult chapters
in geology, and it is to place it in as simple and intelligible a light as the nature of the subject will permit that we
attempt the present Sketch.   Of course, it is not to be
expected that what is often a matter of doubt and difficulty
to the professed geologist can be made easy to the comprehension of the casual inquirer ; but an indication of the
processes of metamorphism can be given, and such indication may lead to a more satisfactory conception of the
character and formation of rocks in general.

One of the most obvious agents in the re-formation of
disintegrated rock-matter is simple mechanical pressure.
As layer after layer of sedimentary matter (mud, clay, sand,
gravel, &c.) accumulates in seas and estuaries, the lower and
earlier are necessarily pressed upon by all those above them,
and thus they are gradually consolidated—muds and clays
into shales, sands into sandstones, and gravels into conglomerates.   This pressure at great depths must be enormous ;
and as it is incessant in its action, we can readily perceive
how rock-substances may be altered in their structure and

texture by this force alone. But during this mechanical pressure chemical actions and reactions are perpetually taking place in all rock-masses, and thus its effects are facilitated and rendered much more perceptible. A bed of peat, for example, may be solidified by compression, but during its chemical passage into coal it undergoes a process of softening and bituminisation which enables mere pressure to act with greater uniformity and effect. In the earth's crust, therefore, wherever chemical transformations take place, or heat is induced by chemical actions, mechanical pressure will act with increased efficiency; and as these are almost everywhere present, we may regard the two combined (that is, pressure and chemical action) as amongst the most important agents in the metamorphism or transformation of rock-matters. And those who have witnessed the effects of the hydraulic press—the conversion of the softest pulp into a solid mass, and the loosest powder into a hard and brittle stone—can have little difficulty in conceiving the myriad-fold greater result of thousands of feet of superincumbent strata, and the continuance of their weight and pressure for unknown ages.

But while all rock-substances are thus continually pressed upon and transformed by those that lie above them, the infiltration of chemical solutions as well as chemical reactions among the particles of their own mass are also materially assisting in changing their composition and texture. The loose shelly sand of the sea-shore is often cemented into a compact conglomerate * by the percolation of rain-water, which, dissolving the limy matter of the shells, diffuses it

* This " littoral concrete," as it has been lithologically termed, may be witnessed among the sand-drifts, and along the shores of many parts of the British Islands. Composed of sand, shells, and pebbles, it is often of stony hardness, and might be mistaken for an older rock, were it not for the imbedded shells, which are all recent, and to be found in the neighbouring seas.

through the mass, as a mortar to bind the incoherent particles into a solid rock. Every rock in the earth's crust—sedimentary and vulcanic—is rendered more compact and crystalline by this process of chemical infiltration. Water is ever permeating this outer shell, and thus solutions of lime, silica, iron, and other mineral and metallic substances are borne hither and thither—here cementing sand and gravel into grits and conglomerates, and there calcifying a sandstone; here silicifying and hardening some earthy limestone, and there converting volcanic dust and ashes into stony tufas and amygdaloids. Besides simply cementing and hardening the masses through which they percolate, these solutions give rise to new chemical actions, by which certain rocks are rendered more crystalline, and altogether changed in their texture. A solution of silica, for example, may permeate a porous sandstone, and by cementing together its particles render it merely harder and more compact; while a similar solution, in passing through an earthy chalk, might form a chemical union with the mass, and convert it into a tough and flinty chertstone. Those acquainted with the changes that can be artificially induced by chemical action, can have little difficulty in conceiving how vast and complex the metamorphoses that may be wrought within the rocky crust by the same agency, and all the more that heat and moisture are there ever present to facilitate its operations. And even where the changes are slow and gradual—so slow as to be almost imperceptible—we may rest assured they are still going forward, and only require time for their perfect development. The recent rocks must necessarily have suffered less metamorphism than the ancient, but still the time and fitting conditions will come when the loose sand will be converted into sandstone, and the sandstone, by further change, into quartzite; the peat be changed into lignite, the lignite into coal, and the bituminous coal into a hard and flameless anthracite.

But whatever the amount of rock-change brought about by chemical agency, it is far less noticeable than that induced by the operations of subterranean heat, whether acting in the dry way, as in volcanoes, or in the wet, as in hot springs and vaporiform exhalations. Whatever be the source of volcanic heat, its effect by contact on all the stratified rocks is at once marked and decisive. Sandstones are frequently converted into crystalline quartzites, shales baked into splintery porcellanite and porcelain jasper, and bituminous coals changed into coke-like anthracites. We every day witness the effects of dry heat in our brick-kilns, coke-ovens, glass-houses, and iron-furnaces; and if the effect of such temporary heat be to melt and bake and calcine, how much more the results of those subterranean fires, that may continue to operate in certain areas unabated for ages! In the earth's crust, then, we may rest assured that volcanic heat has effected, and is still effecting, extensive metamorphism in all rock-masses, and that those lying at great depths and under vast pressure will be affected very differently from those occurring near the surface. But beyond this contact with dry heat there is the permeation of heated water and vapours, which greatly facilitate the dissolution and recombination of mineral matters. As the temperature of the crust increases at the rate of one degree Fahr. for every 60 or 65 feet of descent, the transforming effect of this heat at the depth of a few miles must be enormous; and we can readily conceive (though we cannot witness the operation) how ordinary strata may be converted into crystalline schists and igneous rocks, as the traps and lavas, undergo a process of re-crystallisation, and assume the character of granites. Heat, in whatever form it operates, is, indeed, the great transformer of mineral matter—here baking and hardening, there calcining and melting; here inducing crystallisation, and there, in combination with other forces, filling veins with veinstuffs and metallic ores.

In connection with this deep-seated heat, we may also believe that magnetic currents and crystallisation (to whatever force or forces it may be due) are also effecting a marked metamorphism on all the older and deeper-seated rocks. The earth is in fact a great magnet, through whose exterior crust currents of varying intensity are ever passing; and these, we may rest assured, are actively instrumental in altering the molecular arrangement of rock-masses, and conferring upon them not only new textures, but also new structures, such as cleavage, crumpling, foliation, and other peculiarities which distinguish the slates and schists of the older formations. Experiment has tried to imitate this mode of metamorphism by passing galvanic currents through masses of moistened pottery clay, and the result was re-arrangement of the particles so as to produce *cleavage*, or fissility, such as occurs in roofing-slate, and at the same time a glistening and semi-crystalline texture. Taken in connection with heat, these subtle forces of chemistry, magnetism, and crystallisation, are obviously important modifiers of mineral and metallic matters; and thus, among the oldest rocks, which have been the longest subjected to their influence, we find crystallisation, cleavage, foliation, and kindred phenomena in their greatest intensity. It is to these older rocks that the term METAMORPHIC is generally applied; and though all rock-matters are continually undergoing metamorphism, and in some localities intensely so, yet these old semi-crystalline and highly-altered slates and schists are so intimately associated in character and position that they have been grouped into a system—the " METAMORPHIC SYSTEM " of Systematic Geology.

Summing up our knowledge of this metamorphism or transformation of mineral matter—by which chalk, for instance, can be changed into crystalline marble, and clay into glistening roofing-slate—it may be safely affirmed that the following are the principal agents concerned in its production, even though we may not be always able to determine their precise modes of operation :—1. *Heat by contact*, as when an igneous mass, like lava, indurates, crystallises, or otherwise changes the strata over or through which it passes.   2. *Heat by transmission, conduction, or absorption*, which may also produce metamorphism, according to the temperature of the heated mass, the continuance of the heat, and the conducting power of the strata affected.   3. *Heat by permeation of hot water, steam, and other vapours*, all of which, at great depths, may produce vast changes among the strata, when it is recollected that steam under sufficient pressure may acquire the temperature of molten lava.   4. *Electric and galvanic currents* in the earth's crust, which may, as the experiments of Mr W. Fox and Mr R. Hunt suggest (passing galvanic currents through masses of moistened pottery clay), produce cleavage and semi-crystalline arrangement of particles.   5. *Chemical actions and reactions*, which, both in the dry and moist way, are incessantly producing atomic change, and all the more readily when aided by an increasing temperature among the deeper-seated strata.   6. *Mechanical pressure*, produced by the mere weight of superincumbent strata, and which is obviously concerned in the solidifying, compacting, and hardening of all rock-matter, whether belonging to the superficial or to the deeper-seated formations.   7. *New molecular arrangement by pressure and motion*—a silent but efficient agent of change, as yet little understood, but capable of producing curious alterations in internal structure, especially when accompanied by heat, as we daily see in the manufacture of the metals, glass, and earth-enware. Such, we repeat, are the more general and likely causes of rock-metamorphism; and as it is possible that several of these may be operating within the same locality at the same time, the reader will perceive that no hypo-thesis that limits itself to any one agent can be accepted as sufficient and satisfactory.

It has been already stated, that while every stratum or portion of a stratum, every formation or portion of a formation, may undergo metamorphism, it was to the older and deeper-seated slates and schists that the term *Meta-morphic* was more especially applied — it being chiefly among these that mineral transformations were to be wit-nessed in their greatest intensity. It is true that in many secondary mountains, such as the Alps and Apennines, the stratified rocks are often highly metamorphosed; but this transformation is for the most part partial, some localities remaining little affected, and having all their fossils distinct and legible. Near one centre of vulcanic energy the shales may be converted into dark glistening schists and the limestones into crystalline marbles, while in another cen-tre, and only a few miles distant, the shales and limestones may retain their original sedimentary aspect. Among the primary formations, on the other hand, the metamorphism is general, and whole mountain-ranges and vast tracts of country, like the Scottish Highlands and Scandinavia, are entirely composed of crystalline slates and schists—the clay-slates, mica-schists, chlorite-schists, gneisses, quartzites, marbles, and serpentines of the mineralogist. Among these rocks stratification is indistinct, fossils are obliterated, and the whole succession is massed into one enormous thickness of unknown origin and antiquity. Under this view these rocks have been successively regarded as

"Primary," "Metamorphic," "Azoic," and "Hypozoic,"* or, in other words, as marking the earliest stages of world-history, and before life had begun to make its appearance on our planet.

This view, however, like many others of the earlier geologists, has had to give way to more extended research and newer discovery. Even during the time of the German geologist Werner, a large portion of these primary strata was found to be partially fossiliferous, and separated under the term "Transition," as indicating the passage of the world from an uninhabitable to a habitable state. In our own time these Transition rocks (as will be more fully explained in a subsequent sketch) have been subdivided into "Silurian" and "Cambrian" systems, both of which have yielded abundant forms of life; and still more recently, in the schists and serpentines of the St Lawrence—the equivalents of the old gnarled gneiss-rocks of Scotland and Norway—traces of lowly life have been discovered, this giving rise to the "Laurentian" system — the oldest or earliest strata in which fossils have yet been detected. Or, tabulating the progress, we have first the PRIMARY of the earlier geologists resolved into *Transition* and *Metamorphic*; secondly, the TRANSITION resolved into *Silurian* and *Cambrian*; and, lastly, the METAMORPHIC resolved into *Laurentian* and *Older Crystalline Schists*; thus—

PRIMARY.	TRANSITION.	SILURIAN SYSTEM. CAMBRIAN ,,
	METAMORPHIC.	LAURENTIAN ,, OLDER CRYSTALLINE SCHISTS.

Here, then, it is obvious that what has hitherto been especially regarded as the metamorphic system is merely a

---

* *Primary,* first or earliest formed; *Metamorphic,* changed or transformed in texture; *Azoic,* without life, or destitute of fossil remains; *Hypozoic,* under life, or lying beneath the fossiliferous formations.

vast succession of stratified rocks of true sedimentary origin like any of the later systems, but so altered by mineral transformation that it is only in localities where the metamorphism has been partial and less intense that evidence of their aqueous and fossiliferous nature can possibly be discovered. As the extensive and mountainous tracts in which these schists occur have been but slenderly investigated, subsequent research may yet discover other fossils, and resolve the whole into definite and satisfactory lifesystems. And should such hopes be fulfilled, how inconceivably exalted will our notions of the world's antiquity become—new æons of life and physical activity extending away into a past as vast perhaps in duration as all the later ages that geology has revealed! We can never hope to read aright these earlier pages of world-history, but passages here and there may be recovered, and from these scattered hints we may obtain enough to convince that then as now the physical agencies of nature were ever active and subject to the same laws, and that Life too was present as their accompaniment, though then merely coming into view under the operation of a higher and more complicated law of development.

How then, it may be asked, are we to deal, in a chronological point of view, with the metamorphic rocks in which no fossils have been detected? Shall we describe them as altered Silurians, Cambrians, or Laurentians? or shall we continue to regard them merely as metamorphic strata of unknown age, till some organism has been discovered that may lead to their identification? Where the stratigraphical succession is evident, it may often be convenient to mark them as Silurian, or Cambrian, or Laurentian; but where the succession is doubtful, and no trace of organism has been discovered, it will be much safer still to retain them as mere metamorphic strata. Much error, both in theory and

practice, may arise from adopting an opposite course, while
no inconvenience can result from the use of the term " metamorphic," which merely implies that the rocks under
review have suffered intense mineral change, but advances
no opinion as to their age or chronological co-ordination.
We might colour on our geological maps the schists of the
Scottish Highlands and Scandinavia, as Silurian, Cambrian,
or Laurentian, and support our views by many plausible
arguments, but nothing would be gained by such a course
which is not already secured by the term "metamorphic,"
while the subsequent discovery of their real character would
only be embarrassed by these hypothetical distinctions.
Let us continue, then, to treat these old rocks simply as
" the metamorphic," labouring to reveal their true nature
by the discovery of unobliterated fossils, and encouraged by
the success which, within the last thirty years, has resolved
so much of them into Silurian, Cambrian, and Laurentian
life-systems.

Such is the nature of metamorphism, or that internal
mineral transformation to which all rock-matter in the
earth's crust is incessantly subjected.   Pressure, heat, chemical action, and the other agents above alluded to, are continually solidifying, hardening, and crystallising ; and no
sooner is a sediment laid down, or an igneous mass ejected,
than it begins to be operated upon by one or other of these
forces.   Of course, the latest laid down will have suffered
less change than the older and deeper-seated ; hence it is
chiefly among the latter that metamorphism is to be seen in
its greatest intensity.   It may happen in certain areas, such
as centres of vulcanic activity, that secondary strata may be
as much metamorphosed, or even more so, than any of the
primary ; but still such instances are exceptional, and it
may be safely asserted as a general rule that the older

rocks have undergone the greatest amount of metamorphism
or internal mineral transformation. So great, in many instances, has this change been, that stratification is rendered
indistinct, fossils obliterated, new minerals evolved in the
mass, and the rocks rendered so crystalline and homogeneous that it is difficult to determine whether they are really
of sedimentary or of igneous origin. Normally speaking,
between the most recent and most ancient rocks there will
be every gradation of metamorphism, and it is to this process that the geologist must learn mainly to ascribe the
different aspects and textures that prevail among the rock-
formations of the globe. The atmospheric, aqueous, and
igneous forces may destroy the rocks in one region and
re-form them in another, but metamorphism is the sure and
silent agency by which they are compressed, solidified,
crystallised, and converted into other and other aspects.
No rock—be it shale, sandstone, limestone, coal, ironstone, lava, or greenstone—remains for ever the same. It is
incessantly, however slowly, passing on to other and newer
aspects, and metamorphism is the great wizard-power by
which the transformations are effected. A sandstone, however soft and granular now, will in time, and under the fitting conditions, be converted into a crystalline quartzite, an
earthy limestone into a saccharoid marble, a tender and
bituminous coal into a hard and flameless anthracite, a
porous lava into a close-grained greenstone or basalt, and
a greenstone itself, by farther change, into a large-grained
crystalline granite. When once the operations of metamorphism are fully understood, the geologist has a key to much
that is puzzling and perplexing in his science ; and not till
the efficacy of these is everywhere admitted can he be said
to have adopted the right methods for the solution of his
problems.

As the oldest rocks have undergone the greatest amount

of metamorphism, so it is among them that the geologist often meets with his greatest doubts and difficulties. Defining them as the *metamorphic*, they were at one time regarded as preceding the fossiliferous strata, and as marking a period of the earth's history when life did not exist, and when, of course, its reliquiæ were not expected to be found in the rocky strata. As research extended, fossils were here and there discovered, even in these metamorphic strata; and hence the necessity of arranging them (see Sketch No. 5) according to this new evidence into "systems" and "formations," as had been done with the younger and more fossiliferous strata. But beyond the deepest in which traces of life have been detected, there still lie vast masses of crystalline and granite-looking schists unresolved, and apparently unresolvable, and to such the designation "metamorphic" is still specially applicable, and may ever remain so. In these we find no legible record of world-history: nothing beyond the great facts that they are stratified rocks, and must have been laid down in seas and estuaries like all other sediments, and that millions of ages must have elapsed during their slow conversion from silts and sands and gravels to crystalline schists and granitoid masses. But though this be the present state of geologic knowledge, we dare not, looking at the progress of the past, presume that further discoveries will not be effected. These old rocks may yet tell their tale of life just as the Laurentian and Cambrian have done it before them; for, carry our researches backward as we may, we perceive no traces of a beginning, any more than in the existing operations of nature we see indications of an end.

Once more: let it be clearly understood that *metamorphism* is simply internal mineral change; that under pressure, heat, chemical action, and other kindred forces, all rock-matter in the earth's crust is incessantly undergoing

such transformations ; and that while this crust is alternately wasted and reconstructed by the agents described in
Sketch No. 2, it is mainly from metamorphism its rocks
receive these peculiarities of texture and structure that
confer upon them their distinguishing characteristics and
economic importance.

THE PRIMARY PERIODS.

NOTHING can be clearer than this: if the higher and more
exposed portions of the earth be continuously wasted and
worn down, and the rock-material so wasted be as continuously deposited in the receptacles of lakes and estuaries,
the older deposits (or formations, if you prefer to call
them) will be the deeper, and the newer will be arranged
in order of succession above them. Or put it in this
way: if the sediments of lakes and seas be gradually
converted into solid strata, and now and again upheaved
into dry land by vulcanic forces, the oldest, generally
speaking, will be the most consolidated and altered. Or
again, if the sediments of lakes and seas contain less or
more of the remains of plants and animals that have been
either drifted from the land or entombed in the areas where
they flourished, then the latest enclosed remains will be

the least altered and most resemble those still living in the district. These propositions are so evident that it would be little else than waste of time to attempt a more detailed explanation.

We have, then, in the crust of the earth some formations very recent and others very ancient—so recent that the formative processes are still in action, and so ancient that it requires all the appliances of modern science to say how they were formed, and to determine the nature of the fossil organisms they contain. The great plain of China and the delta of the Mississippi, for instance, are of recent growth, and indeed still in course of accumulation; and by a parity of reasoning we may readily perceive that the prairies of North America and the pampas of South America, though somewhat older, are still of comparatively recent formation. Everything connected with these accumulations—mineral composition or imbedded organisms—has an air of recentness about it, and is readily intelligible; but it is different with the sandstones and limestones that lie beneath, whose structure has been altered, and whose fossils have been mineralised and rendered obscure. These sandstones and shales and limestones were no doubt at one time loose sands and clays and calcareous muds, but pressure, chemical action, and other metamorphic forces, continued through long ages, have converted them into solid strata, and this conversion has been intensified, for the most part, according to their relative dates. As with the stratified or sedimentary formations, so also with the igneous. Such hills as Hecla, Etna, and Teneriffe readily reveal their history; but the Alps, the Pyrenees, and the Grampians, whose internal fires have been long since extinguished, whose rocks have become more crystalline, whose heights have been repeatedly beneath the waters, and from which all the loose volcanic matters have been washed and worn away, present a very different aspect, and require all the ingenuity of science to interpret their history.

This relative antiquity of rock-deposits may be still more clearly shown by the formations of our own islands. The carse-lands of the Tay and Forth and the fens of Lincolnshire are recent alluvia, and younger than the blue clays and gravels over which they are spread ; the stratified sands and clays and gravels in the neighbourhood of London are younger than the chalk and greensands that lie beneath ; and these chalk-hills and greensands of Kent and Surrey are newer than the calcareous sandstones and limestones of Portland on which they repose. Again, the oolites or roe-stones of Portland are younger than the underlying red sandstones and marls of Cheshire, and these newer than the coals and ironstones of Lancashire that are spread out beneath. The coals of Lancashire, Northumberland, and Fife are not so old as the underlying red sandstones and flagstones of Forfarshire ; and these again are much younger than the still deeper slates and crystalline schists of the Scottish Highlands. Nor is it superposition alone that proves this relative antiquity. Mineral structure and texture become more intensified with age ; fossil organisms become more obscure, and the further we descend in time the wider the divergence from the genera and species that now people the globe. Everything in the earth's crust speaks to this relative antiquity—to this old, older, oldest ; and it is the object of the present Sketch to describe the earlier of these formations, and to depict, as far as Geology can, the aspects of the periods when they were gradually accumulating in the primeval waters.

Before entering on this description, however, it may render matters more intelligible to state, and what indeed has been already detailed in No. 1 of these Sketches, that modern geologists arrange the earth's crust into the following *formations*, or, referring to their fossils, into the following *life-periods* and *rock-systems* :—

Life-Periods.	Rock-Systems.
CAINOZOIC=RECENT LIFE.	{ Quaternary or Recent. { Tertiary.
MESOZOIC=MIDDLE LIFE.	{ Cretaceous or Chalk. { Oolitic or Jurassic. { Triassic or Upper New Red.
PALÆOZOIC=ANCIENT LIFE.	{ Permian or Lower New Red. { Carboniferous or Coal System. { Old Red Sandstone and Devonian. { Silurian.
EOZOIC=DAWN LIFE.	{ Cambrian. { Laurentian.

This arrangement is, of course, temporary or provisional, and may be altered as geologists become more intimately acquainted with rocks of other lands ; but in the mean time it expresses our knowledge of the succession that prevails among the stratified formations, and may be received as the great chronological stages of the world's history. The technical terms are founded partly on mineral, partly on geographical, and partly on fossil distinctions, and may with a little exertion become intelligible to the least scientific. Thus "Cretaceous" and "Old Red Sandstone" refer to the most prevalent rocks in these systems ; "Laurentian" and "Cambrian" to districts where the systems are largely or typically displayed ; and "Cainozoic" and "Palæozoic" to the comparative recentness and antiquity of the fossil remains. A uniform system of nomenclature might have been preferable, but human knowledge progresses by slow degrees, and its terms and technicalities must be viewed as mere provisional expedients towards this advancement. Under the existing nomenclature geology has made bold and rapid progress, and any attempt to revolutionise, unless through the gradual increase of wider knowledge, would tend to obstruct rather than facilitate. It is to the oldest of these so-called systems that we now direct attention; to the conditions under which they seem to have been deposited; and to the kind of life that appears to have peopled the lands and waters of their respective periods. We have classed them as the "Primary Periods," because there is really a greater similarity between their rocks and fossils than there is between the rocks and fossils of any subsequent periods; because, so far as we know, their strata are all truly marine deposits; and further, because their fossil forms belong (speaking in general terms) to *invertebrate* types—zoophytes, shell-fish, crustacea, &c.—and are specially characterised by the absence of the *vertebrates*— the fishes, reptiles, birds, and mammals.

Beginning with the LAURENTIAN, which term has been employed by Sir William Logan of the Canadian Geological Survey, to designate the highly crystalline strata which belong especially to the valley of the St Lawrence, and constitute the great bulk of the Laurentide mountains, we may take Sir William's own description of the rocks which compose this oldest and deepest of sedimentary formations. "The rocks of this system," he says, "are almost without exception ancient sedimentary strata which have become highly crystalline. They have been very much disturbed, and form ranges of hills having a direction nearly north-east and south-west, rising to the height of 2000 or 3000 feet, and even higher. The rocks of this formation are the most ancient known on the American continent, and correspond probably to the oldest gneiss of Finland and Scandinavia, and to some similar rocks in the north of Scotland. They consist, in great part, of crystalline schists (chiefly gneissoid or hornblende), associated with felspars, quartzites, and limestones, and are largely broken up by granites, syenites, and

diorites, which form important intrusive masses. Among
the economic minerals of the formation, the ores of iron are
the most important, and are generally found associated with
limestones." Interpreting Sir William's technicalities, it
may be stated for the comprehension of the general reader
that this old Laurentian formation, which is of vast thickness (some 30,000 feet or thereby), consists essentially of
hard and crystalline strata like the gneiss, mica-schists,
quartz-rocks, and marbles of the Scottish Highlands, or more,
perhaps, like the still harder and more granitic-looking
schists of the Scandinavian mountains. There are no sandstones, or shales, or limestones in the proper sense of the term.
All these have been converted, long ages ago, by heat, pressure, and chemical action, into sparkling crystalline rocks;
lines and layers of stratification are obscure and often altogether obliterated; veins and eruptive masses are frequent;
and altogether the whole formation wears the aspect of a
vast and venerable antiquity.

That the Laurentian system, like other stratified systems,
was deposited in the form of sands, gravels, clays, muds,
and other loose sediments, is beyond all question. Nature
has no other mode of procedure. What is wasted from the
lands is transferred to the waters; nothing is lost. It may
change its form or place, but it is still in existence; and this
incessant round of waste and reconstruction, as shown in a
former Sketch (No. 2), is the ordained order of the universe. What a wonderful metamorphism these primeval
sediments have undergone! Not mud, nor sand, nor
gravel; not shale, nor sandstone, nor conglomerate; but
glistening slates and crystalline schists — sandstones converted into flinty quartz-rocks, and limestones into variegated serpentines. And in the midst of all this metamorphism, the fossil organisms seem to have shared the same
fate; for we cannot think of waters of deposition without

associating with them some forms of life, however lowly. And yet, till within the last two years, the search for traces of life in these primeval rocks was considered visionary, and *azoic* and *hypozoic*, or "lifeless" and "under-all-life," were the technicalities by which they were known. *Nil desperandum*, however, should be the geologist's motto, and especially of those who believe as we do that Life on this globe was coeval with the stratified rocks, and that the conditions which permitted the deposition of ordinary sediments must have been favourable at the same time to the manifestation of some form or other of vitality. And so it happens that traces of lowly organisation have recently been detected in these old Laurentian rocks—in the serpentinous limestones of Canada—thus holding out the hope that the primeval rocks of other regions will yet yield similar traces, and prove that the earliest waters were tenanted by their own forms of life, and were gladdened by the manifestations of sentient existence.

Indeed similar organisms have already been detected in the old (and probably contemporaneous) serpentines of Ireland and marbles of Bohemia; and as one form of life generally indicates the existence of another upon which it preyed or was in turn preyed upon, we may shortly expect important additions to this discovery. Even while we write (1866), tracks and burrows, supposed to be those of worms, have been discovered in a somewhat higher zone, and described by Principal Dawson of Montreal. The organism discovered, *Eozoön Canadense*, or "Dawn-animal of Canada," belongs to the lowest forms of life—lower than the infusory animalcules, and even still lower than the sponges. It is one of the Foraminifera, mere animated specks, which have nevertheless the power of secreting lime from the waters, and enveloping themselves in elegant and variously-formed cases. These calcareous cases are mere microscopic points, and yet they are perforated by numerous pores (*foramina*, hence the name of the order), through which the creature procures its food and holds intercourse with the outer world of waters. Individually minute, they live in colonies, and only become conspicuous by their aggregations, which in the instance of the eozoon vary from a few inches to a foot or two in diameter. The white calcareous mud which covers so much of the Atlantic sea-bed is a similar foraminiferal accumulation; so is the nummulitic limestone which stretches eastward in a great zone through Europe and Asia; so also is a large proportion of the chalk-hills of England; and so backward in time through other limestones, till we reach the oldest and earliest Laurentian marble. With a little manipulation, the organisms constituting the existing sea-muds or the chalk are readily revealed; but the old eozoon has to be polished, cut into microscopic slices, and treated with acids before the peculiarities of its structure can be rendered intelligible. So great is the change produced by the mineralisation of ages.* Strange that the minutest of organisms should be capable of piling up such stupendous rock-masses; strange and suggestive

---

* We are aware some geologists have called in question the organic nature of this Canadian *eozoön*—regarding it merely as a peculiar mineral structure mimetic of the organic, examples of such simulative structures being well known in other formations. The majority of competent observers, however, maintain its organic nature; and from a conversation we had, in the autumn of 1865, with Principal Dawson of Montreal, who has examined the rocks *in situ*, with their unobliterated lines of deposition and layers of organic growth (brought out more clearly by weathering), we share the conviction that the *Eozoön Canadense*, whatever be its zoological affinities, is of animal and not of mineral aggregation. To the practised eye external appearances are often conclusive of organic structure, and these, in the present case, had been observed and accepted before the microscope was called into court to complete the evidence. Those interested in this matter may refer to the papers by Dr Carpenter, Professor King, and others, in the Journals of the Geological Society for 1865 and 1866.

that the lowest forms of life should be the first and earliest to reveal itself to the lenses of the palæontologist!

Turning next to the CAMBRIAN system, so termed by Professor Sedgwick, because well developed in the region of Wales (the ancient Cambria), we find it for the most part made up of crystalline schists and slates, hard siliceous grits, and altered limestones. Like the Laurentian, its strata have undergone extensive metamorphism or internal mineral change, and like the Laurentian, too, it is frequently inter- sected by veins and eruptive masses of granite, and fels- pathic greenstone. On the whole, however, its sedimentary character is much more apparent ; its micaceous schists are interstratified with grits and sandstones, and its slates often earthy and more distinctly laminated. Taking it in the mass of its 15,000 or 20,000 feet, it has altogether a more recent aspect than the Laurentian, and bears, not only in its sandstones and shaly slates, but in its imbedded pebbles, ripple-marks, and tracks of marine worms, more decided evidence of its aqueous origin. Of course, like other forma- tions, the Cambrian will vary in composition in different regions, sometimes being more slaty, and at others more schistose and crystalline ; but slaty beds, micaceous flag- stones, gritty sandstones, and limestones more or less crys- talline, may be taken as the normal aggregate.

When we turn to the fossils of the system we find them also more numerous and intelligible than those of the Lauren- tian, partly no doubt from the less metamorphism the beds have undergone, but chiefly, perhaps, from that advancing development of life we are accustomed to associate with the newer and newer formations. At all events, instead of a single organism, as in the Laurentian system, we now find, if not a numerous, at least a fair array of zoophytes, echino- derms, shell-fish, annelids, and crustacea. The species are certainly not the highest of their respective orders, neither are they exactly the lowest; but, making ample allowance for the defects of our present information, it may be safely asserted that the *fauna* or animal-life of the Cambrian period is altogether one of a lowly character, and that of the *flora* or plant-life we know nothing beyond a few indistinct impressions of algæ or sea-weeds. Even the little we now know of the Cambrian flora and fauna was altogether unknown twenty years ago, and it is chiefly since 1846, and more especially since 1859, that its fossiliferous character has been fairly established.

Here then, as in the case of the Laurentian system, we have a long period of the earth's history—so long that 18,000 or 20,000 feet of sediment was accumulated in certain parts of the ocean—and of which we know nothing beyond what is recorded by these marine strata and the fossils they contain. We have no glimpse of the land from which these sediments were worn and wasted, yet there must have been broad lands to supply this waste and rivers to transport it. We are utterly ignorant of the plant-life and animal-life—if terrestrial life then existed—by which these lands were peopled. We know nothing of the disposition of sea and land as compared with the existing continents and seas. All that we clearly perceive is the existence of these sediment-receiving oceans, with their scattered sea-weeds, zoophytes, shell-fish, and crustacea, and only faintly and at distant intervals their sandy and muddy shores, in which annelids bored and left their burrows, and over which crustacea tracked and left their traces. Little as this knowledge may seem, it is everything compared with the beliefs of our ancestors; it is a great deal compared with what was known even by the last generation; and it holds out the encouragement that another generation, by following in the right path, will arrive at a fuller insight into the physical and vital aspects of this primary period.

The last of the primary systems which form the subjects of our Sketch is the SILURIAN, so named by Sir Roderick Murchison, because well displayed and first examined by him in that border country between England and Wales which in ancient times was inhabited by the Silures. Rocks of Silurian age have been found in almost every region of the globe—in Central and Northern Europe, largely in both Americas, and in Australia—and though they differ much in their mineral composition, some being more crystalline and slaty than others, still on the whole there is a wonderful similarity among them, both in their lithological and fossil aspects. Occasionally they are so metamorphosed as to be undistinguishable from the crystalline schists of the Cambrian and Laurentian; but, generally speaking, their sedimentary character is abundantly apparent in the numerous alternations of sandstones, slaty shales, and limestones, and we see in these strata, with their fossil corals, shells, and crustacea, the clearest evidence of deep and widespreading seas. Altogether the geological record becomes more legible, and we can form some notion of the earth's terraqueous conditions during the long and gradual deposition of the Silurian sediments. We say long and gradual deposition, for in our own islands these strata are from 20,000 to 30,000 feet in thickness, embracing numerous alternations of rock-material, and repeated removals and renewals of genera and species of animals.

In these Silurian strata we perceive limestones formed of coral-reefs and calcareous debris, slates, and slaty shales arising from the deeper sea-muds, and sandstones, grits, and conglomerates composed of the sands and pebbly shingle of the shallower waters. Here and there through the mass we

find interstratified overflows of lava and showers of volcanic ashes, indicating that then, as now, the vulcanic forces
were active in their work of upheaval and eruption. How
wonderfully well these old rocks have retained the record
of their history! Here shore-formed sandstones pattered
by crustacean feet and riddled with worm-burrows; there
limestones formed of coral-growth and shell-debris in the
deeper waters: here slaty shales composed of the slimy
mud of the stiller depths, and replete with zoophytes; and
there at intervals vast sheets of tuff and porphyry that
had been showered abroad as ashes or ejected as lava from
submarine volcanoes. Occasionally, in the pebbles of the
conglomerates, we catch glimpses of the kind of rocks that
formed the lands from which these sediments were transported, and from drifted clubmoss-like twigs in the shales
we know that these lands were clothed in some degree with
a vegetation, however lowly.

Such is the tale told of the Silurian epoch by its own rock-
formations; but the history receives a deeper and livelier
interest when we come to consider the number and variety
of organisms imbedded throughout the system. It is true
that these are of lowly orders and wholly invertebrate, if
we except a few scattered fish-remains found in the very uppermost beds, and by many regarded as belonging more properly to the Old Red Sandstone; but lowly as they are, they
occur in vast exuberance and variety, and mark a marvellous
progress in life-development compared with what is known
of the Cambrian and Laurentian. Sea-weeds and drifted
twigs and spores of clubmoss-like land-plants is all we
know of the Silurian *flora;* but certain beds of anthracite
and anthracitous shales favour the idea that plant-life was
then more exuberant than has yet been detected. In its
fauna or animal-life we have foraminiferal organisms,
sponges, corals, polyzoa, or aggregate animals like the sea-

mats and sea-pens (graptolites, &c.); shell-fish of every order, bivalve and univalve, deep-sea and shore dwellers; radiate animals, like the encrinites and the star-fishes; sea-worms in their tracks and burrows; and crustaceans, chiefly trilobites and eurypterites, having some resemblance and affinity to the existing limulus or king-crab. These organisms are not found indiscriminately throughout the system, but vary in number and distribution according to the kind of stratum in which they are imbedded—every zoophyte and shell-fish preferring a certain kind of sea-bottom; and according as they occur in the lower, middle, or upper portion of the system—the upper being the more prolific and characterised by the higher species. Numerous and varied as they are, they are (with the exception of the obscure land - twigs) exclusively marine; and if we regard the uppermost beds, with their fish-remains, as the base of the Old Red Sandstone, they are entirely invertebrate,* and mark, so far as our present evidence goes, the close of a long primary cycle during which vitality was gradually evolving, in a fixed and definite order, from lower to higher manifestations.

As economic repositories these primary systems are, in some regions, of considerable importance; less, however, for their rock-products than for the metalliferous veins by

---

* The reader must guard against the idea that there are any sharp lines of demarcation between the so-called *Systems* of geologists. The life of certain estuaries and seas may no doubt be brought to a close by some sudden catastrophe, but such breaks are merely local, and do not affect the general life-arrangements of the globe. When we speak, therefore, of the Silurian as "marking the close of a long invertebrate period," it is not meant to be asserted that there were absolutely no fishes during the deposition of the uppermost Silurians, but simply *that the Primary Periods as a whole were characterised by their want of vertebrate remains,* and that the strata in which fish-remains do occur may be regarded, without detriment to the science, either as uppermost Silurian or as lowermost Old Red Sandstone.

which they are traversed. The vast bulk of their strata being gneisses and mica-schists, are little fitted for architecture; but their limestones, marbles, and serpentines are often of great beauty and much sought after for ornamental purposes; while their fine-grained cleavable clay-slates afford abundant material for roofing and other multifarious appliances. Indeed, it is chiefly in these old metamorphic rocks that serpentines, variegated marbles, and roofing-slates occur, the mineral transformations to which they have been subjected giving to the former those varied shades of colouring and figure, and to the latter that fissility or cleavage, for which they are prized. But if the rocks of these systems are of comparatively little value, their metalliferous veins are rich and numerous—gold, silver, tin, copper, antimony, manganese, iron, and other metals being abundant in most primary districts, either in the veins themselves or in the debris that has been worn and washed from them during the course of ages. These primary rocks constitute, indeed, the bulk of all our older hill-ranges, and it is only in them that the slow processes of chemical deposition have yet elaborated on a grand scale the metallic ores, and the vein-stuffs with which these ores are usually associated. It is true that veins and ores occur in some of the younger formations, but not in the same variety, nor with the same richness and persistence, as those that belong to the primary and more highly metamorphic strata. Hence, it may be remarked, the importance of geological investigation in colonial and newly-discovered countries; the determination of their formations being tantamount to a declaration of their mineral and metallic wealth, or, in other words, their natural fitness for mechanical and commercial development.

Such are the great primary periods of world-history—the Laurentian, the Cambrian, and Silurian. Geologists may name and arrange them as they may, but the great fact stands unquestioned, that according to our present knowledge they form the deepest or earliest of the fossiliferous systems. Other fossiliferous strata still deeper and older may be discovered as geology extends her survey of the world's crust, but in the mean time such formations are unknown. Whole systems of strata, deeper and older than the Laurentian, may have been obliterated by metamorphic changes, but such reasoning is altogether futile. In earth-history as in human history we must make our chronological arrangements according to our knowledge, and the furthest verge to which we have pushed our investigations must necessarily stand as the practical, though provisional, commencement. And, after all, there is something so like a beginning in these old Laurentian rocks, with their lowly eozoa, that we feel, if not at the confines, at least nearing the confines, of the possible in geological history. Observe, it matters nothing to this history though the earth had swung for untold cycles as an incandescent but gradually cooling mass. We have no means of inductively determining such a condition, and we are bound in reason to commence our history with the earliest operations we can discover in the crust. The earliest of these operations is the laying down of sediments as nature is now laying down sediments; and when we find these imbedding forms among the very lowest of organised existence and nothing higher, and find that in after ages the higher gradually make their appearance in orderly succession, it would be abandoning all logical guidance if we did not arrive at the conviction that we were approaching, in these Laurentian quartz-rocks, schists, and serpentines, the commencement of the existing ordainings of our planet. It is true, that Dr Dawson's discovery of worm-burrows in the upper Laurentians, if further corroborated, would carry the inference

G

of Life-beginnings still lower than the lowermost Laurentians, and into strata which, in the absence of fossil remains, we are in the habit of designating " metamorphic ;" but then be it observed that the distance in time between the upper and lower Laurentians is very vast, and that the carrying of lowly life-forms even beyond the Laurentian epoch does not materially affect the conclusion. It is merely believed we are approaching, not asserted we have reached the ultimate limit of vitality.

Of course everything connected with these investigations is as yet dim and difficult. We know next to nothing of the areas occupied by Laurentian and Cambrian strata, and can only sketch in very general terms the boundaries of the Silurian. We know nothing of the lands from which the Laurentian and Cambrian sediments were borne, and can only dimly indicate the direction from which some of the Silurian were transported. The vast thickness of these formations, the frequent alternations of their strata, and the fineness of the sediments, imply long lapses of time— but how long it were in vain, without some standard of comparison, to inquire ; and even though we could give the time numerical expression in years and centuries, we could form no adequate conception of its immensity. All that we know for certain is, that the earth in these primeval periods had its seas and continents—seas in which sediments were deposited, and lands from which the material must have been borne. With the exception of a few Silurian stems and seed-vessels, we are in utter ignorance of the life and aspects of these lands ; and are merely left to infer the existence of other forms of life from the presence of those we have determined. Of the tenantry of the seas we know that they were few and simple at first, but, as time rolled on, newer and higher types made their appearance, and this not only in greater numerical abundance, but in greater specific variety. No doubt the record is very imperfect, and we cannot for a moment suppose that all the forms of life have been revealed by the few scattered patches geologists have examined. Still, there must be a meaning in these lowliest forms of life coming first, in their comparative scantiness in the earliest formations, in the gradual appearance of higher and higher forms, and in their increasing abundance and specific variety; and that meaning, if we would interpret without bias or predilection, is surely this—that in these old formations we are approaching—if indeed we have not already reached—the dawning of life and the first orderings of that system of progression that still prevails, and is carrying the present along with it.

Such, once more, is the faint and fragmentary history of the earth's primal periods—dim and shadowy loomings of undiscovered lands; glimpses of broad seas in which the lowlier forms of life spread and increased in numbers, and rose by some great creative law through newer species to higher and higher orders. Fragmentary as the history may seem, it is truly marvellous that science, dealing with such obscure materials, should have been able, within little more than a quarter of a century, to arrive at such satisfactory conclusions. Within the memory of living geologists these Silurian, Cambrian, and Laurentian systems were the "Transition," "Metamorphic," "Hypozoic" (under life), and "Azoic" (void of life) rocks of the systematists; and now each and all of them have yielded their life-forms—proving the great fact, that there is no known period of the earth's history when life, in some form or other, did not exist, and leading to the belief that the manifestation of life on our planet was coeval with the earliest of the stratified formations. How wonderful the interest with which these chips and fragments can invest the history of the past! Like the prehistoric relics of the shell-mounds, the lake-dwellings, the caves, and the river-gravels, they may lead to no very definite or connected view of the aspects and history of the periods to which they relate; but to science they are invaluable as proving what has been there, and as indicating the design and method of creation. Where everything is so obscure the faintest glimmer becomes an unspeakable boon—a certainty of itself wherefrom to predicate, and the strongest incentive to further investigation.

VEINS—THEIR NATURE AND ORIGIN.

IT has been stated in the preceding Sketch, that of all the
rocks in the earth's crust, the primary are those most abundantly traversed by veins. As these veins are the great
repositories of the metallic ores, it may be useful at this
stage briefly to explain what veins are, how they occur, and
what the general character of their contents. This information may lead to a better comprehension of much that will
be subsequently stated, at the same time that it is valuable
knowledge of itself, and belongs to one of the most interesting departments of modern geology.

It was mentioned under Vulcanicity, that in all volcanic
areas the solid crust was more or less rent and fissured—
these rents either radiating from some centre of eruption,

or. running parallel to each other according to the most
prevalent direction of the earthquake convulsions. These
fractures will vary, of course, from mere cracks to yawning
chasms many feet in width, and will descend at all inclinations—some sloping downwards at a low angle, and others
sinking perpendicularly, or nearly so. Now it requires no
great effort of the understanding to perceive that these
fissures, in course of time, will be filled either by matter
washed in from the earth's surface, or by volcanic and
mineral substances injected from the interior. Such rents
in the stratified rocks, when thus filled up by lava, by
greenstone, or by basalt, are generally known as " dykes,"
the igneous matter rising up like a wall through the strata
on either side. On the other hand, when they are filled
by sparry or crystalline minerals, and these intermingled
more or less with metallic ores—the slow and gradual depositions of chemical agency—they are usually distinguished
as " Veins," from their traversing and ramifying through
the crust like the *veins* through the animal system. But
the veins that were formed at one period may be cut
through or crossed by others of a later era, and thus in
many districts there is a network, as it were, of veins,
crossing and intercrossing in a very complicated manner.
As might be expected, too, the original veins may contain
one kind of mineral or metal, and the cross-veins another
kind, and hence the greater richness, as well as complexity,
of many metalliferous regions.

Understanding, then, that " dykes" consist of unproduc-
tive rock-matter, and that " veins " are always less or more
metalliferous, it may be stated as a fact, that the latter
occur most abundantly, and naturally so, in the primary
formations. These are the rocks that have suffered most
from igneous convulsions, and these also are the rocks
among which metamorphism and chemical agency have had

longest time to bring about internal change and fill with crystalline and metalliferous depositions. There are, no doubt, productive veins in later formations, such as the lead and silver bearing veins of the Carboniferous limestone; but these by no means occur in the same richness and variety as those of the primary strata. Wherever, then, we have extensive developments of primary rocks and mountains, as in Wales and Cornwall, Scandinavia, the Ourals, the Andes and Mexican Sierras, there also we may expect a corresponding development of metalliferous veins —gold, silver, tin, copper, and the like—of varying age and richness, according to certain laws the order and governance of which geology is yet unable to indicate. The formation and accumulation of rocks is in most instances a slow and gradual process, but the segregation and deposition of metalliferous matter is still slower, and thus we may look upon the veins of the primary strata as of high antiquity, though necessarily younger than the rocks they traverse. Occurring most abundantly in the older formations, and very rarely in those of secondary or tertiary date, it is to the veins and veinstones of primary regions that the following remarks will be more especially directed.

Defining a vein as a rent or fissure in the earth's crust which has been subsequently filled up by infiltrations of mineral and metallic matter, it must be obvious that veins will be of various widths and of various inclinations. Productive veins seldom exceed a few feet in width, and it is rare to find them beyond fifteen or twenty; but their inclinations are at all angles—some descending almost perpendicularly, and others sloping downwards by very easy stages. The bounding rocks on either side form the *cheeks* or *walls* of a vein; the mineral matter of which it is composed, the *vein-stuff,* *matrix,* or *gangue;* and the metallic ore is distributed through the matrix in *ribs,*

pockets, *nests*, *strings*, and *plates*, according to the manner
and abundance of its occurrence. The vein-stuff is usually
arranged in layers from the walls inwards, the centre being
generally occupied by a rib of ore, though not unfrequently hollow and lined with crystals. The whole matrix
has thus a striped or veined appearance, the stripes running up and down or parallel to the cheeks or containing
walls. It must be obvious from this description that a
vein is something very distinct from the rocks through
which it passes. If it pass through igneous rocks, its
stripes and colours contrast very strongly with the dark
uniform hues of these masses; if through sedimentary
rocks, its upward and downward course through their strata
at once arrests the attention; and, generally speaking, its
sparry or crystalline texture is sufficient to define its thickness and direction.   Passing from below upwards, and
frequently ramifying, crossing, and intercrossing in many
directions, they look indeed like the veins in vegetable and
animal structures, and hence their appropriate and expres-
sive designation.

The sparry matter which forms the bulk of the vein-
stone or matrix consists for the most part of quartz, carbonate and sulphate of lime, carbonate and sulphate of
baryta, or of alternations and admixtures of these—the ore
occupying a subordinate part in ribs, strings, nests, and
pockets. These vein-stuffs seem to have been deposited first
on the cheeks, and then coating after coating towards the
centre, which is either solid like the rest of the matrix, or
hollow, as if there had been a deficiency of filling matter;
and, in such cases, the cavity is lined with crystals shoot-
ing and pointing inwards. Having had room to assume
their independent forms, the crystals in these cavities are
often of great beauty, and it is usually from such vein-
spaces that the mineralogist obtains his rarest treasures.

In some instances the vein-stuff consists of a single substance in repeated coatings, such as quartz or carbonate of lime; in other instances it is made up of alternating layers of two substances, such as lime and baryta; and in many cases it consists of a seemingly capricious admixture of several ingredients. As with the vein-stuffs so with the enclosed metallic ores—some veins containing only one metal, others two or more metals, and in such cases there is usually a curious and persistent connection, as lead with silver, copper with tin, iron with manganese, and gold with platinum. Besides this curious connection of metal with metal, there is often an observed relation between certain ores and certain vein-stuffs, as gold in quartz, lead in carbonates and fluates of lime, &c.; and these relations when carefully noted are often of great practical value to the mineral explorer. What has caused these curious alternations of vein-stuffs and connections of certain metals science cannot in the mean time determine, but it observes the modes in which they occur, and this, when accurately done, is always a step towards the solution. Whatever the causes, they have operated not always in filling the veins and fissures merely, but often in impregnating the adjacent rocks through which the veins pass, and thus the cheeks or walls are occasionally worked for the strings and plates of the ores they contain.

But though unable to explain the relations of certain ores to certain vein-stuffs, geology has amassed a great deal of observation on the subject, as well as on the prevalent direction or *strike* which veins take in certain localities. The ascertaining of this direction is all-important; for while the main veins of a district containing one kind of metal are found to strike always in one direction—easterly and westerly, for instance—the secondary or cross veins running northerly and southerly are almost certain to be the repositories of other kinds of ore. As the main veins or "lodes" are thus intersected by others of more recent date, they will also be more or less displaced, and the ascertaining of these facts is ever of the highest moment to the mining industry of a district, to say nothing of its importance to correct geological deduction. The mapping of these directions and the colouring of the primary and secondary veins according to the kind of ores they contain, is a work to be done by the mining-engineer for each respective district; though it is now well known, that according to the rocks of a country and the age and direction of its hills, so also are the general strikes of the veins and the nature of their metalliferous contents. The direction of veins and the nature of their contents reveal a chronology and order within their respective areas as much as the stratified systems do, and a knowledge of these relations is alike of scientific and economic importance.

At the present time attention has been less directed to the direction of veins and the corresponding nature of their contents than to the modes of their formation and their subsequent infiltrations. At one time igneous action was called in by hypothesists, not only to produce the original fissures, but to accomplish also the subsequent fillings-in; and, in fact, to perform the most opposite and contradictory functions. Now, however, while it is agreed that the rents and fissures were originally produced by, and owe their linear directions to, vulcanic commotions, it is the general and growing opinion that their subsequent fillings-in with sparry and metallic matters are due to the infiltration and deposition of chemical solutions. Heated waters and vapours, or *hydro-thermal action*, as it has been termed, has been the main agent in dissolving, carrying upwards, and re-depositing the mineral and metallic matters—that is, the sparry vein-stuffs as well as the crystallised ores. In fact, similar spars and ores can be produced in the laboratory by chemical and electrical means, and all the more certainly that heat is present to facilitate the operation. What can be simulated on the small scale by art we may readily believe to be producible by nature on the large, and thus to chemical and electro-chemical agency are now generally attributed the formation of vein-stuffs and their associated ores. We have spoken of the greater efficiency of heated waters in dissolving and holding in solution mineral and metallic matters, but the effects of water in general (whether hot or cold) must not be overlooked, it being the grand medium through which the contents of veins have been conveyed to their present positions. It is this belief in chemical solution and re-deposition which distinguishes the modern theories of mineral veins from the older views of sublimation through igneous or plutonic action.

Whatever may have been the immediate agent of infiltration and deposition, one thing is certain, that veins, and especially metalliferous ones, are most abundant in areas that have been long subjected to igneous action. There is the closest connection, and necessarily so, between the two phenomena, as it is only in convulsed districts that rents and fissures can occur, and in such districts also that such fissures have the greatest chance of being most rapidly filled with mineral and metallic precipitates. Wherever, therefore, there are old hill-ranges and primary areas that have been repeatedly subjected to subterranean forces, there we may expect veins and cross-veins—each set representing a long course of time, and being for the most part filled with its own special ores and vein-stuffs. All portions of a hill-range may not be rich alike, for there are certain centres in which the producing forces seem to have acted with greater intensity, or at all events to have been longest continued, and it is in these that the greatest variety of veins and cross-veins occur, and from these also that the greatest variety of metallic ores are to be obtained. What the law that has determined the greater richness of certain districts, science cannot as yet give the slightest indication, any more than it can tell why certain areas that were once convulsed with igneous activity have long since been cold and silent. All that can be done in the mean time is simply to note the facts, and these, when correctly recorded, become of the greatest importance to industrial operations, as they will one day or other do to scientific deduction.

The importance of correct information on all that relates to metalliferous veins and deposits cannot be too highly valued, and especially in countries like Britain, that depend so much upon the metals for their mechanical, manufacturing, and commercial greatness. Whether as a medium of exchange, for the fabrication of implements and the construction of machines, or merely for objects of luxury and ornament, the metals are all-important; and whatever tends to certainty and facility in obtaining their ores is deserving of a nation's encouragement. Without the metals there cannot, indeed, be high and substantial progress in civilisation; and in modern times a nation's place may be safely indicated by the facilities she has of obtaining them. Her hills may be bleak and barren, and little fitted for the amenities of agriculture; yet beneath that poor and rugged surface may lie mines of untold wealth, and the readiest means of manufacturing and commercial greatness. And such is usually the contrast that presents itself in mining and metalliferous districts. Cold and retentive clays, ungenial moorlands and uplands, are too often characteristic of coal tracts, as witness those of Northumberland, Durham, and Lanarkshire; while cliffs, and scars, and bleak unapproachable ridges, are the common concomitants of metalliferous regions, as those of Derbyshire, Wales, and Cornwall.

The great value of primary districts lies, as already mentioned, in their metalliferous lodes and veins, or in the stream-drifts that have been weathered and worn in course of ages from the cliffs and precipices above. The vein lies in the solid rock, and must be mined with great labour and outlay; the stream-drift, on the other hand, is but the water-borne debris from the veins above, and demands merely sorting and washing. The stream-work is the ready and primitive method of obtaining the ores and metals; the mine is the laborious but more certain appliance of modern times and modern requirements. In conducting a stream-work, little more is needed than manual labour and care; in managing a mine, mechanical appliances, engineering skill, and correct geological deduction are indispensable at every stage of the undertaking.

Such is a brief, and necessarily sketchy, outline of the nature and origin of veins and vein-stuffs. The rents or fissures originally produced by subterranean convulsion are subsequently filled by infiltrations of mineral and metallic matter, and thence become the *veins* which seem to ramify and reticulate through the earth's crust like the veins through vegetable and animal structures. A fissure may be produced in an instant by earthquake convulsion, but ages may pass before it be completely filled by sparry minerals and metallic ores—the slow depositions from aqueous percolation and solution. As water is ever percolating the earth's crust, so it is ever dissolving from one part and re-depositing in another; and this power of dissolving is no doubt greatly augmented by heat, just as rapidity of precipitation and crystallisation may be facilitated by electro-magnetic currents which are incessantly traversing the

rocky framework.   In this way the fissure becomes a vein ;
and as each set of veins has a fixed direction, and is
charged with its own peculiar vein-stuff and metal, the
ascertaining of these directions and peculiarities is at once
of the highest scientific and economic importance.   As the
original fissures are produced by volcanic convulsions, and
the subsequent fillings-in by slow and gradual precipitation
from solution, veins will occur most abundantly, of course,
in districts that have been longest subjected to those
agents ; and such tracts are necessarily those occupied by
primary and transition strata.   In these districts, bleak
and barren and inhospitable, the mining industry of the
world is chiefly situated, their subterranean wealth compensating, and often more than compensating, for their want
of agricultural fertility and amenity.   As the metals are indispensable to mechanical, manufacturing, and commercial
progress, so they are generally regarded as powerful auxiliaries of civilisation ; hence the importance of all that attaches, scientifically and industrially, to their history, their
modes of occurrence, the means of obtaining them, and the
processes of reducing them from their ores and associated
vein-stuffs.

FOSSILS—THEIR NATURE AND ARRANGE-
MENT.

WHATEVER may have been the meaning which our forefathers attached to the term *fossil* (Lat. *fossĭlis,* dug up),
every man and woman of ordinary intelligence now understands that it refers to the remains of plants and animals
found in the crust of the earth, and more or less petrified
or converted into stony matter. Where these remains—
whether trunks, branches, or leaves, bones, teeth, or shells—
occur in recent and superficial accumulations, they appear
little altered in texture, and are usually looked upon as
sub-fossil, or only partially fossil; but when they are imbedded in the older and harder strata, the stony conversion
is in general complete, and they are then regarded as true
fossils or petrifactions. Wherever they are found their
history excites a lively interest; and minds altogether un-
attracted by the physical record of the earth are often

excited to enthusiasm in the search of its organic memorials. Their study is, in fact, a kind of archæology—an antiquarianism like that which attaches to ruins and burial-mounds, but of a broader and more marvellous description. To the older geologists their occurrence was a riddle, and few considered them as other than mere accidents or *lusus naturæ;* but to the modern geologist they are replete with information of the world's past, revealing to him the kind of life that peopled its lands and waters during the successive stages of its history, and, by inference, the geographical conditions under which they flourished and declined.   The recognition of their nature and importance has thrown a new and higher interest round geology ; and where the study of mere rocks and minerals formerly shed an uncertain glimmer, the science of fossils has cast the light of sure and satisfactory information.   This science of fossils, or *Palæontology*, as it is technically termed (Gr. *palaios*, ancient; *onta*, beings; and *logos*, discourse or reasoning), is now, indeed, one of the main sections of geology; for if geology be world-history, that history can never be written without a knowledge of the plants and animals that have successively peopled the earth, as well as of the external conditions which the nature of these plants and animals alone can indicate.   It is necessary, then, that the student of popular geology should know what fossils really are, the various states in which they occur, and the manner in which they can be arranged according to the classifications of the botanist and geologist.   To these subjects we devote the present Sketch,—premising that Palæontology may be technically subdivided into *Palæophytology*, or the science of fossil plants, and *Palæozoology*, or the science of fossil animals ; though, for all ordinary purposes, the general term is sufficiently comprehensive and intelligible.

Like other things in the history of the earth, the nature

and occurrence of fossils will be best explained by an appeal to the existing operations of nature. If we stand by the side of a river, and especially when it is in flood, we perceive that the current is continually bearing onward vegetable and animal debris, and that this debris is gradually entombed among the mud, sand, and gravel which the river deposits in the lake, estuary, or sea into which it discharges its waters. As with one river, so with every rill and river that traverses the terrestrial surface — each is carrying down the spoils of the land in a state more or less fragmentary, and burying them in the silt, where, excluded from atmospheric decay, they appear in the first stages of fossilisation. As with land plants and animals, so with those of lakes and seas; they die and are imbedded in sediments where they lived and grew, or are drifted by tides and currents to some distant locality. This process is ever going forward in every region—tropical, temperate, and arctic; and as each region is characterised by its own special flora and fauna, the imbedded remains will indicate to future observers the external conditions under which they grew and were deposited. The plants and animals entombed in the estuary of the Amazon must differ from those deposited in the delta of the Mississippi, and these again from those preserved in the mud-islands of the Niger, the Ganges, and other Old World rivers. The shells, crustacea, and fishes that die and sink in the sediments of tropical seas differ widely from those of temperate waters, and these again as widely from the fauna of the Arctic and Antarctic Oceans. As it is now, so it must have been in all former ages, and thus the fossils of the stratified formations become the only clue to the geographical conditions of the areas in which they were deposited, and of the regions from which they were derived.

Nor is it geographical or climatic condition alone of which

H

these fossil relics bear evidence. Every family of plants has its own peculiar station—the waters, the marsh, the plain, the upland, or the shingly desert; and every family of animals its own special habitat—the forest, the open plain, the shallow lake, the sandy or muddy shore, or the greater ocean-depths. As these families are regulated now, so the palæontologist presumes they were governed in former epochs, and thus by a critical study of his fossils he arrives at a more vivid picture of the past, and can associate with each order and family the general features of their physical surroundings. From his knowledge of the present he rises to a true conception of the past, and from his acquaintance with the existing he can indicate with something like certainty the habitats and distribution of the extinct. It is true there will be occasional comminglings of terrestrial and aquatic remains, of fresh-water and marine, just as now the spoils of the land may be mingled with those of the estuary, and those of the river with those of the ocean; but in general such commixtures are limited, and do not obliterate the broader characteristics of the formations in which they occur. Here and there the record may be complicated; it is never equivocal or disguised.

Nor is it mere habitat and distribution the palæontologist can thus arrive at; but habit and function are also determinable by the requisite anatomical skill. The forelimb to swim, the forelimb to walk, the forelimb to run, the forelimb to seize, and the forelimb to fly, are each stamped by its own essential characteristics, just as the herbivorous, the carnivorous, and the insectivorous teeth are; and thus the competent palæontologist is enabled to recall not only the physical surroundings of his fossil flora and fauna, but their forms and functions—presenting a picture of the world's past like that which the geographer presents of its existing phenomena. There are few things, indeed, which science has greater cause to boast of than this determination and restoration of fossil forms. From a few stray chips and fragments to reveal the nature of the plant or animal to which these fragments belong, or from a few scattered bones and teeth to reconstruct the form of the creature and indicate its habits and functions, is, in truth, the triumph of modern anatomy. " *Every organised being,*" says the immortal Cuvier, "*forms a whole, a single circumscribed system, the parts of which mutually correspond and concur to the same definite action by a reciprocal reaction. None of these parts can change without the others also changing, and consequently each part, taken separately, indicates and gives all the others.*" In this truth lies the fundamental law of the co-relation of parts, the discovery of which enabled the great French anatomist to effect his wonderful restorations of the mammals of the Paris Basin, and the enunciation of which has ever since thrown the light of hope and of certainty over the toilsome labours of the palæontologist. To him fossils became, as they have been eloquently and appropriately termed, the MEDALS OF CREATION; "for as an accomplished numismatist, even when the inscription of an ancient and unknown coin is illegible, can from the half-obliterated effigy, and from the style of art, determine with precision the people by whom and the period when it was struck; so in like manner the geologist can decipher these natural memorials, interpret the hieroglyphics with which they are inscribed, and, from apparently the most insignificant relics, trace the history of beings of whom no other records are extant, and ascertain the forms and habits of unknown types of organisation, whose races were swept from the face of the earth thousands of ages before the creation of man and the creatures which are his contemporaries."[*]

* Mantell's ' Medals of Creation,' vol. i. p. 17.

Of course, all plants and animals are not preservable alike, nor are the same organisms, though occurring in the same formation, always found in the same state of preservation.   Plants and animals that have been exposed to atmospheric decay before entombment will be less perfect than those that have been suddenly and thoroughly imbedded.   The harder parts of plants—roots, stems, leathery leaves, and nut-like fruits—will run a better chance of preservation than the soft and succulent portions.   Corals, shells, crusts, bones, teeth, scutes, and scales, will be preserved when all the softer parts of the animals to which they belong have entirely disappeared.   The harder and more massive portions of a skeleton will resist when the softer and more slender have fallen to decay.   The dense and thoroughly ossified bones of an old animal will endure where the spongy and unanchylosed members of a young one fall asunder and perish.   Ferns, mosses, and resinous pines will resist maceration when other plants will totally disappear.   All things considered, aquatic animals run a better chance of preservation than terrestrial ones; and the bulkier land-mammals and amphibia than the birds and insects.   Gregarious animals, too, are likely to be found in greater abundance than those living isolated and solitary — the catastrophe (earthquake, land-flood, or windstorm) which would destroy only a few of the latter, overwhelming the former by hundreds of thousands.   In this way shoals of fishes may be suddenly suffocated by submarine exhalations, shell-beds buried beneath obnoxious sediments, herds of ruminants borne from the land by floods, and clouds of insects swept into the sea by wind-storms. The inconceivable numbers in which fossils are sometimes found crowded into very limited areas would seem to point to such accidents for their entombment—their perfection, indiscriminate aggregation, and individual positions, all indicating some sudden death-catastrophe.

The nature of the sediments must also exercise a marked influence in the number and perfection of the imbedded fossils. Loose and porous sands will be less preservative than impervious clays and muds, and heterogeneous silts than calcareous sediments. In this way minute organisms may totally disappear, and larger ones be preserved in a mutilated and fragmentary form. Every one acquainted with the nature of our peat-mosses, with the sands, clays, and marls that fill up our ancient lakes, and with the sands, gravels, and silts now accumulating in our estuaries and sea-creeks, must have witnessed the different preservative effects of these sediments ; how solid and dense the bones are in one, and how spongy and rotten they appear in another ; how hard and firm the shells are in one, and how soft and friable they occur in another ; and how sharp and clear every external marking is retained in one matrix, and how wasted and obscure it becomes in another. As with these recent accumulations, so with the strata of the older formations ; some are destitute of organic remains, which must at one time have been imbedded in them in abundance, while in others they are so imperfectly preserved as to be of little or no value to the palæontologist. The various ways in which plants and animals may be imbedded and preserved in sediments being so obvious, the different preservative effects of different sediments being also apparent, the reason why some forms should occur more abundantly than others being generally discoverable, and the evidences which such plants and animals afford of the geographical conditions under which they flourished being admitted, let us now inquire into the processes by which they are lapidified, or converted into stony matter.

Generally speaking, in recent accumulations, such as sand-silt, peat-moss, and the like, the remains of plants and animals are found little altered. The more volatile matters are expelled from the plant, and the more perishable integuments and softer tissues of the animal have perished. Roots, trunks, branches, and the harder fruits, if excluded from the action of the atmosphere, become darker, denser, and assume a peaty aspect. In like manner, bones, horns, teeth, shells, and crusts, lose a portion of their animal matter, and become denser and heavier through some slight absorption of mineral ingredients. But in all the older formations the vegetable, if not converted into coal, is thoroughly lapidified—that is, changed by a slow chemical process into flint, ironstone, limestone, or sandstone, as the case may be—and merely retains its organic form ; while animal remains undergo a similar conversion, and are recognisable only through their individual forms and textures, which continue unchanged and persistent. There is nothing more marvellous than this process of petrifaction : particle after particle as the organic matter disappears, so particle after particle the mineral matter takes its place, and this so delicately that scarcely a cell or fibre is ever ruptured or displaced ! Of course where the mineral solutions percolating the earth are so numerous there will be great variety of petrifactions, some being calcareous or limy, some siliceous or flinty, some ferruginous or irony, and others bituminous or coaly. But in whatever state they may occur the process seems the same—namely, a gradual dissipation, through decay, of the organic atoms, and a gradual substitution, through permeation, of the mineral or inorganic. Great differences will also arise from the chemical nature of the organisms themselves—wood, bone, horn, teeth, shells, crusts, and corals, each having its own composition, and possessing its own power of resisting decay.   Not only so, but as the percolation of mineral solutions through the earth's crust is incessant, what is deposited at one time may be dissolved at another and a new substance substituted in its place, or no new substance may be substituted, and merely the hollow mould of the organism left to prove that it once was there.   For instance, a shell or coral, which consists of animal-formed carbonate of lime, may be converted into sparry mineral carbonate ; or this may be dissolved and carried away, leaving merely a hollow mould with every ridge and line and pore impressed on the containing matrix; or this mould may be refilled with siliceous matter, and the shell or coral then present itself as a flint, with every pore and ridge and wrinkle as delicately perfect as on the original organism the day it was imbedded.   This perfection of preservation is often, indeed, truly marvellous.   We have seen the faceted eyes of trilobites as perfect in form as when they received the rays of light through Silurian waters ; carboniferous univalves with their colour-bands still unobliterated ; internal casts of productæ with their muscular apparatus displayed in a style of legibility to which no anatomical preparation could approach ; and ink-bags of cuttle-fishes so little changed as to furnish the pigments for their own portraiture.

Of course, the consideration of these percolations, dissolutions and substitutions, involves many intricate questions in chemistry; but enough has been stated to inform the general reader that fossils may occur in many different conditions —as stony conversions, as moulds, as casts, or as mere impressions on the matrix in which they are entombed.   But in whatever state they may occur, there is usually sufficient left either in general form, in external character, or in internal texture, to enable the palæontologist to determine their

places, or at all events to approximate to their places, in the
vegetable and animal schemes.   These determinations, indeed, constitute the main duty of palæontology; and when
one considers how widely scattered fossils usually are, how
sorely mutilated and fragmentary they often appear, and
that they are largely the chance findings of quarrymen and
miners, it is truly marvellous how much of reliable world-
history the science, within little more than half a century,
has been able to reveal.   It is true there is still very much
to be done, and perhaps more in fossil botany than in fossil
zoology, inasmuch as vegetable organisms are less perfectly
preserved than animal, and because the classifications of the
botanist are mainly founded on the flowers and leaves—
portions which of all others are the most evanescent and
perishable.   Notwithstanding these difficulties, the palæontologist finds that all his fossils belong to the same great
scheme of life with existing plants and animals, and he is
therefore restricted to the classifications that have been de-
vised by the botanist and zoologist.   Species, and genera,
and families, and even what are called orders, may have
become extinct, and others in the course of creation may
have taken their places; still the great Scheme of Vitality
has ever been evolved according to a fixed and determinate
plan, and in harmony with this plan, and to the best of
their knowledge, botanist, zoologist, and palæontologist must
endeavour to conform their systematic arrangements.

In speaking of fossil plants, therefore, the palæophytologist
adopts the usual classification of the botanist, placing where
he can his fossils under their proper genera and orders, and
where he cannot, assigning to them a provisional place next
to the genus or order to which they bear the greatest resemblance.   In this course he uses the same botanical terms,
and employs the same botanical phraseology, and these may

be rendered intelligible in a general way by the study of the annexed tabulations :—

The Vegetable Kingdom may be arranged into two grand divisions—the CELLULAR and VASCULAR ; and these again, according to their modes of growth and reproduction, into the following groups and classes :—

I. CELLULAR—Without regular vessels, but composed of fibres which sometimes cross and interlace each other. The *Confervæ* (green scum-like aquatic growths), the *Lichens* (which incrust stones and decaying trees), the *Fungi* (or mushroom tribe), and the *Algæ* (or sea-weeds), belong to this division. In some of these families there are no apparent seed-organs. From their mode of growth—viz., sprout-like increase of the same organ—they are known as ·THALLOGENS or AMPHIGENS, and constitute the lowest orders of vegetation.

II. VASCULAR—With vessels which form organs of nutrition and reproduction. According to the arrangement of these organs, vascular plants have been grouped into two great divisions—CRYPTOGAMIC (no visible seed-organs), and PHANEROGAMIC (apparent flowers or seed-organs). These have been further subdivided into the following classes—ascending from the lower to the higher forms :—

1. CRYPTOGAMS—Without flowers, and with no visible seed-organs. To this class belong the *mosses, equisetums, ferns,* and *lycopodiums.* It embraces many fossil forms allied to these families. From their mode of growth—viz., increase at the top or growing point—they are known as ACROGENS.

2. PHANEROGAMIC MONOCOTYLEDONS—Flowering plants with one cotyledon or seed-lobe. This class comprises the *water-lilies, lilies, aloes, rushes, grasses, canes,* and *palms.* In allusion to their growth by increase within, they are termed ENDOGENS.

3. PHANEROGAMIC GYMNOSPERMS—This class, as the name indicates, is furnished with flowers, but has naked seeds. It embraces the *cycadeæ* or cycas and zamia tribe, and the *coniferæ* or firs and pines. In allusion to their naked seeds, these plants are also known GYMNOGENS.

4. PHANEROGAMIC DICOTYLEDONS—Flowering plants with two cotyledons or seed-lobes. This class embraces all forest trees and shrubs—the *compositæ, leguminosæ, umbelliferæ, cruciferæ,* and other similar orders. None of the other families of plants have the true woody structure, except the *coniferæ* or firs, which seem to hold an intermediate place between monocotyledons and dicotyledons ; but the wood of these is readily distinguished from true dicotyledonous wood. From their mode of growth—increase by external rings or layers—they are termed EXOGENS.

SCHEME OF VEGETABLE CLASSIFICATION.

SPERMOCARPS *(Seed-Fruits.)*

- **ANGIOSPERMS** *(Enclosed Seeds.)*
  - EXOGENS......DICOTYLEDONS......Trees, Shrubs, Herbs. *(Increase from without.)* *(Two Seed-Lobes.)*
  - ENDOGENS......MONOCOTYLEDONS......Grasses, Sedges, Palms. *(from within.)* *(One Seed-Lobe.)*
- **GYMNOSPERMS** *(Naked Seeds.)*
  - GYMNOGENS......POLYCOTYLEDONS......Cycads and Conifers. *(from without.)* *(Many Seed-Lobes.)*

SPOROCARPS... *(Spore-Fruits.)*

- **ANGIOSPORES** *(Enclosed Spores.)* ... ACROGENS.... *(from the point.)*
  - SPOROGAMS......Clubmosses, Lycopods. *(Spore-Growths.)*
  - THALLOGAMS......Ferns and Horsetails. *(Leaf-Growths.)*
  - AXOGAMS......Mosses and Liverworts. *(Centre-Growths.)*
- **GYMNOSPORES...AMPHIGENS...** *(Naked Spores.)* *(from all sides.)*
  - HYDROPHYTES......Algæ and Confervæ. *(Water-Growths.)*
  - AËROPHYTES......Lichens. *(Air-Growths.)*
  - HYSTEROPHYTES......Fungi or Mushrooms. *(Ultimate or Lowest Growths.)*

Subdividing still further, according to their most marked characteristics, whether external or internal, the botanist arranges all the known forms of Vegetable Life into some 300 orders, upwards of 9000 genera, and about 120,000 species. As the most of these distinctions, however, are founded on the form and connection of the flower, fruit, and leaf—organs which rarely or never occur in connection in a fossil state—the palæontologist is guided in the main by the great structural distinctions above adverted to, and not unfrequently by the simple but unsatisfactory test of "general resemblance." The flower and the organ of fructification may have perished, but still the form and venation of the leaf, the external sculpturing of the bark, the disposition of the leaves and branches, and the general mode of growth, may be preserved; and from these, as well as from a microscopic examination of the lapidified tissues, the palæophytologist can for the most part determine, or at all events approximate to the determination of, his fossil twigs and fragments. So certain, indeed, are the determinations of the microscope, that where good sections can be procured, the competent observer rarely or never fails to establish the great order to which the organism belongs; and this, considering the difficulties surrounding palæophytology, is a triumph of no mean description.

As the palæophytologist, in arranging his fossil organisms, is guided by the classification of the botanist, so the palæozoologist follows, as closely as the nature of his objects will permit, the systematic schemes of the zoologist. And in this he has altogether an easier task, inasmuch as animal organisms, from their less destructible nature, are in general more perfectly and legibly preserved. The horny and calcareous structures of zoophytes, corals, shells, crusts of crustacea, calcareous tubes of annelids, chitonous wing-sheaths of insects, and the like, are well-known instances of this comparative indestructibility among the invertebrata; while bony scales and scutes, horns, bones, and teeth are still more familiar examples, perhaps, among the vertebrata.   In this way, partly by external form, partly by internal structure, and partly by the great anatomical law of the co-relation of parts, the palæozoologist is enabled to arrive at determinations more satisfactory, on the whole, than those of the palæophytologist.   Species, genera, families, and even whole orders, may be extinct; but, comparing his organisms with the existing, he finds their nearest affinities, and assigns them their place in the systematic arrangements of the zoologist.   For this purpose the subjoined scheme is in general sufficient, and its study in this place will greatly facilitate the reader's comprehension of what may be subsequently stated respecting the fossils of the different formations :—

SCHEME OF ANIMAL CLASSIFICATION.

VERTEBRATA,

Or animals with backbone and bony skeleton, and comprehending

MAMMALIA, AVES, REPTILIA, and PISCES.

I. MAMMALIA, or *Sucklers;* subdivided into Placental and Aplacental.

1. PLACENTAL, bringing forth mature young.

BIMANA (*Two-handed*)—Man.
QUADRUMANA (*Four-handed*)—Monkeys, Apes, Lemurs.
CHEIROPTERA (*Hand-winged*)—Bats, Vampire-bats, Fox-bats.
INSECTIVORA (*Insect-eaters*)—Mole, Shrew, Hedgehog, Banxring.
CARNIVORA (*Flesh-eaters*)—Dog, Wolf, Tiger, Lion, Badger, Bear.
PINNIPEDIA (*Fin-footed*)—Seals, Walrus.
RODENTIA (*Gnawers*)—Hare, Beaver, Rat, Squirrel, Porcupine.
EDENTATA (*Toothless*)—Ant-eater, Armadillo, Pangolin, Sloth.
RUMINANTIA *Cud-chewers*)—Camel, Llama, Deer, Goat, Sheep, Ox.
SOLIDUNGULA (*Solid-hoofs*)—Horse, Ass, Zebra, Quagga.
PACHYDERMATA (*Thick-skins*)—Elephant, Hippopotamus, Rhinoceros.
CETACEA (*Whales*)—Whale, Porpoise, Dolphin, Lamantin.

2. APLACENTAL, bringing forth immature young.

MARSUPIALIA (*Pouched*)—Kangaroo, Opossum, Pouched Wolf.

MONOTREMATA (*One-vented*)—Ornithorhynchus, Porcupine-ant-eaters.

II. AVES, or BIRDS.

RAPTORES (*Seizers*)—Eagles, Falcons, Hawks, Owls, Vultures.

INSESSORES (*Perchers*)—Jays, Crows, Finches, Sparrows, Thrushes.

SCANSORES (*Climbers*)—Woodpeckers, Parrots, Parroqueets, Cockatoos.

COLUMBÆ (*Pigeons*)—Common Dove, Turtle Dove, Ground Dove.

RASORES (*Scrapers*)—Barnfowl, Partridge, Grouse, Pheasant.

CURSORES (*Runners*)—Ostrich, Emu, Apteryx.

GRALLATORES (*Waders*)—Rails, Storks, Cranes, Herons.

NATATORES (*Swimmers*)—Divers, Gulls, Ducks, Pelicans.

III. REPTILIA, subdivided into Reptiles Proper and Batrachians.

1. REPTILES PROPER.

CHELONIA (*Tortoises*)—Turtles, Tortoises.

LORICATA (*Covered with Scutes*)—Crocodile, Gavial, Alligator.

SAURIA (*Lizards*)—Lizard, Iguana, Chameleon.

OPHIDIA (*Serpents*)—Vipers, Snakes, Boas, Pythons.

2. BATRACHIANS, or FROGS.

ANOURA (*Tail-less*)—Toad, Frog, Tree-Frog.

URODELA (*Tailed*)—Siron, Triton, Salamander.

APODA (*Footless*)—Lepidosiren, Blindworm.

IV. PISCES, or FISHES.

SELACHIA (*Cartilaginous*)—Chimæra, Sharks, Sawfish, Rays.

GANOIDEA (*Enamel-scales*)—Amia, Bony-pike, Sturgeon.

TELEOSTIA (*Perfect-bones*)—Eels, Salmon, Herring, Cod, Pike.

CYCLOSTOMATA (*Circle-mouths*)—Lamprey.

LEPTOCARDIA (*Slender-hearts*)—Amphioxus.

INVERTEBRATA,

Or animals void of backbone and bony skeleton, and comprehending

ARTICULATA, MOLLUSCA, RADIATA, and PROTOZOA.

I. ARTICULATA, subdivided into Articulates and Vermes.

1. ARTICULATA, or Jointed Animals Proper.

INSECTA (*Insects*)—Beetles, Butterflies, Flies, Bees.

MYRIAPODA (*Many-feet*)—Scolopendra, Centipedes.

ARACHNIDA (*Spiders*)—Spiders, Scorpions, Mites.
CRUSTACEA (*Crust-clad*)—Crayfish, Crabs, Shrimps, Woodlice.
CIRRHOPODA (*Curl-feet*)—Acorn-shells, Barnacles.

2. VERMES, or Worms Proper.

ANNELIDA (*Small-rings*)—Lobworm, and almost all the marine worms.
ROTIFERA (*Wheel-bearers*)—Rotifers, Hydatina.
GEPHYRIA (*Intermediates—urchin-like*)—Sipunculus, Echinurus.
LUMBRICINA (*Earth-worms*)—Earth-worms, Nais.
HIRUDINEI (*Leeches*)—Leeches, Branchellion.
TURBELLARIA (*Turbellaries*)—Planaria, Ribbon-worms.
HELMINTHES (*Gut-worms*)—Intestinal worms.

II. MOLLUSCA, subdivided into Mollusca and Molluscoida.

1. MOLLUSCA, or Shell-fish Proper.

CEPHALOPODA (*Head-footed*)—Cuttle-fish, Octopus, Calamary, Nautilus.
PTEROPODA (*Wing-footed*)—Clio, Hyalæa.
GASTEROPODA (*Belly-footed*)—Snails, Slugs, Whelks, Cowries.
ACEPHALA (*Headless*)—Oysters, Mussels, Cockles, Shipworms.
BRACHIOPODA (*Arm-footed*)—Terebratula, Lingula.

2. MOLLUSCOIDA, or Mollusc-like Animals.

TUNICATA (*Coated, but Shell-less*)—{ Biphora, Simple and Compound Ascidians.
POLYZOA (*Compound animals*)
or } Flustra, Eschara, Plumatella, &c.
BRYOZOA (*Moss-like animals*)

III. RADIATA, or ZOOPHYTES—Ray-like Animals.

ECHINODERMATA (*Urchin-skinned*)—Sea-urchins, Star-fishes.
ACALEPHÆ (*Sea-nettles*)—Jelly-fish, Beroes.
POLYPI (*Many-feet*)—Coral animals, Sea-anemones, Hydras.

IV. PROTOZOA, or LOWEST-LIFE—Globular Animals.

INFUSORIA (*Infusories*)—Monads, Volvoces, Vorticella.
PORIFERA (*Pore-bearers*)—Sponges, Fresh-water Sponges.
RHIZOPODA (*Root-footed*)—Amœba, Polythalamia (Foraminifera).

Such are the leading facts connected with the nature, history, and arrangement of fossils. Much more might have been stated respecting the positions in which they occur and the geographical conditions which they thereby indicate, and a more detailed account might have been given of the chemical theories of petrifaction; but enough has been mentioned to convey to the reader a fair idea of what fossils really are,* how they are formed, and how indispensable their study is to the right interpretation of the history of our planet. With this preliminary knowledge he will peruse with greater zest and intelligence the Sketches that follow; and be better able to trace through the ascending stages of time that plan of vital development, the elucidation of which has conferred on modern geology its highest interest and most enduring attraction.

Nor is it Geology alone that has benefited by the discoveries of the palæontologist. Botany and Zoology have also acquired new interest, and the whole study of Life assumed a broader and more philosophical bearing. Restricted to existing forms, the biologist was often perplexed by anomalies he could not solve, and for want of connections he could not trace; but now that Palæontology has revealed its myriad forms, and exhibited a Scheme of Life ever ramifying, yet ever interblending in its remotest ramifications, a clearer and steadier light has been thrown across the path of his investigations. Even to the ordinary observer of nature, how much more exalted the conceptions of life which the science of palæontology imparts! How

---

* Fossils are sometimes arranged into the following classes :—*First*, the actual substance; *secondly*, the substance replaced by other substances; *thirdly*, the cast or mould of the substance—and this may be either of the hard or of the soft substance; and, *fourthly*, those fossils which are now generally called physiological impressions, such as footprints, being certain evidence of the animal having been there. Under whatever class they may be arranged, their preservation will depend partly on their own composition, partly on the nature of the stratum in which they are imbedded, and partly on the chemical changes to which that stratum may have been subsequently subjected. The investigation of these particulars, however, belongs more to the professed palæontologist than to the readers of general geology.

marvellous that, numerous and varied as are the plants an
animals of the present day, they form but the merest fra
tion of those that have successively adorned the earth
surface, each succeeding age being characterised by its ow
special forms, ascending and still ascending in variety an
complexity, yet all interwoven into one grand and ha
monious Life-system !

THE OLD RED SANDSTONE.

THERE is the ring of antiquity in the very title of our sub-ject, and yet old as the Old Red Sandstone may be, it is younger by unnumbered ages than the Laurentian, the Cambrian, and the Silurian. These earlier sediments were converted into hard and crystalline strata, and upheaved into dry land, long before it was deposited, and in many instances they formed the hills and precipices from which its materials were derived. Its place in the earth's chrono-logy will be seen at a glance from the accompanying tabu-lation; but its interest as a formation arises less from its antiquity than from the fact of its being the first in which vertebrate remains decidedly occur, and from the circum-stance that its history has been rendered classical by the labours of some of our leading geologists. Hugh Miller's 'Old Red Sandstone,' Agassiz' Monograph of its fossil fishes, the investigations of De la Beche, Murchison, Pan-der, Huxley, the American States surveyors, and others, have all contributed to this result; and during the last

I

twenty years there are few systems whose names, at least
have been more familiar to the ordinary reader.　But since
Agassiz elaborated his monograph, and Miller penned his
sketches, more extensive information has been obtained
and it is the object of the present chapter to display that
newer knowledge in an intelligible and attractive form.

Arranging the rock-formations of the crust in chronolo
gical order, it will be seen that the Old Red Sandstone
holds a middle place among the palæozoic or primeval :—

Quaternary or Recent, . . ⎱	CAINOZOIC.
Tertiary, ... ⎰	(*Recent Life.*)
Cretaceous or Chalk, . . ⎱	MESOZOIC.
Oolitic or Jurassic, . . ⎰	(*Middle Life.*)
Triassic—Upper New Red Sandstone,	
Permian—Lower New Red Sandstone,	
Carboniferous, ...	PALÆOZOIC.
Old Red Sandstone and Devonian,	(*Ancient Life.*)
Silurian, ...	
Cambrian, ... ⎱	EOZOIC.
Laurentian, . . . ⎰	(*Dawn Life.*)

It does not belong to the very oldest, whose rocks have
been rendered crystalline by metamorphism, and whose
fossils have been sorely obliterated, but it is still very an
cient, and hence the interest that attaches to its old-world
forms, the outlines of which and their habits of life the
pen of the palæontologist can for the most part restore
The composition and origin of its strata are, generally
speaking, of easy determination.　Conglomerates that were
once pebble and shingle beaches ; sandstones and flagstones
resulting from shore-formed sands; concretionary and coral
line limestones chiefly of animal origin ; and shales and
marlstones, the consolidated muds of the deeper waters
Here and there we have bituminous shales, partly of animal
and partly of vegetable impregnation ; and at still wider in

tervals thin seams of coal, like those of Gaspé in Canada, which seemed to have resulted from the growth and drift of terrestrial vegetation. Wherever it occurs its sedimentary character is sufficiently apparent, and though frequently intersected by dykes and eruptive masses of basalt and felstone, its stratified arrangement is never wholly obliterated. The interstratifications of volcanic ash and igneous overflows observable in the Silurian system, and so frequent in the Carboniferous, are rarely witnessed in connection with the Old Red Sandstone, as if the period, in the north of Europe at least, had been one of comparative internal quiescence. The system occupies considerable areas in Europe, Asia, Africa, and both Americas, and is chiefly of marine formation, though in some districts the total absence of shells and corals would lead to the inference of freshwater conditions.*

We have said that the system occupies extensive areas both in the Old and New World, and as no two rivers carry down the same kind of debris, and no two seas receive exactly the same kind of sediments, there will be considerable diversity in the character of its rocks—that is, in colour, composition, and arrangement. Not only so, but as the system is often of great thickness (12,000 feet or more), there had been oscillations of the crust or new distributions of sea and land during the long period of its deposition, and thus its lower, middle, and upper portions differ even in the same region, and sometimes lie unconformably upon each other. It is for this reason that geologists speak of

---

* If the Old Red Sandstone of Scotland be of marine origin, it seems inexplicable why no sea-shell, coral, or other zoophyte should have yet been detected in any of its strata. Numerous as its fishes and crustacea undoubtedly are, and gigantic as some of them may appear, they may have been inhabitants of estuaries or fresh-water seas ; and though the general belief leans towards oceanic conditions, we are still without unmistakable proofs to support it.

the "Lower Old Red," the "Middle Old Red," and the
"Upper Old Red,"—each series differing not only in the
composition of its strata, but in the character of its fossi
contents.   But whatever its variations, there is, throughou
Europe at least, a marked prevalence of reddish-coloured
sandstones and slaty shales; hence the name "Red" i
allusion to this colour, and the term "Old" because i
lies beneath the coal-measures, and in contradistinction t
another series of red sandstones (the New Red) that lie
above them.   The system is also frequently termed th
"Devonian," because a portion of it is well developed i
Devonshire—a term chiefly introduced by Sir Roderic
Murchison, to harmonise with his geographical nomencl
ture of Silurian and Permian.   So much for name an
mineral composition; let us now try to catch a glimpse (
the physical conditions under which it was deposited, an
the kind of life that peopled the land and waters.

Beyond a few scattered indications of the ancient distri
butions of sea and land, geology can obtain no more.   Or
formation is so frequently overlaid by portions of later fo
mations; so many portions have also been removed t
waste and denudation; and perhaps still greater expans
are hidden by the ocean, which covers nearly three-fourtl
of the earth's known surface, that we can merely indica
by disconnected patches the seas in which they were d
posited.   In the case of the Old Red Sandstone, whic
occupies considerable areas both in the Old and New World
we cannot trace either the extent or configuration of i
seas, but we catch occasional glimpses of their shores in t
conglomerates which must have formed their pebbly beache
and in the worm-trails and burrows, the crustacean track
the rain-prints, and sun-cracks on the surface of the san
stones which must have then spread out as shallow ai
alternately exposed sands.   Strange revelations these of t

olden sea-shore!—the ripple of the receding tide, the winding trail of the shell-fish, the burrow and sand-cast of the
sea-worm, the patter of crustacean feet, the pittings of the
rain-shower, and the irregular shrinkage cracks of the sun-
baked shore-mud. And yet, as surely as these phenomena
are witnessed on the muds of existing sea-creeks, so surely
were they impressed on the shores of the Old Red Sandstone, were dried and hardened by the sun, covered over by
newer sediments, and thus preserved through all time as
evidences that nature's operations have been going forward
much in the same way from the remotest of periods. But
clear as these physical evidences are of the nature of the
Old Red sea-shore, there are facts connected with the great
extent and thickness of the pebbly (we may say bouldery)
conglomerates that are not so easy of explanation. We
know that in many parts of the world there are vast pebbly and shingly beaches, and that in some instances the
rounded blocks are hundreds of pounds in weight; but
there is something so peculiar in the aggregation of the Old
Red conglomerates, with their striated pebbles, their irregular imbeddings of fine-grained sandstones and the like,
that they suggest the idea of masses floated and packed up
by shore-ice, and perhaps to some such condition their enormous accumulations may yet be ascribed.* Be this as it

* Several years ago we appended the following note to a chapter on
the Old Red Sandstone (' Past and Present Life of the Globe'), and see
no reason yet to change our opinion :—Whoever has examined the bouldery conglomerates of the Scottish Old Red, with their large irregular
blocks, their peculiar unassorted aggregation, the nature of the cementing matrix, and the frequent "nestings" or interlaminated patches of
fine argillaceous sandstone, must have had suggested to his mind the
idea of ice-action. And this notion must have been strengthened when
he turned to the sandstones, and found them imbedding angular fragments of rock, shale, and even clay, which could scarcely have suffered
transport unless enclosed in drifting ice-floes. The paucity of terrestrial
life in certain areas seems also a further corroboration of the idea of
glacial influences—a hypothesis which seems at first sight extremely

may, its usual sandstones, flagstones, slaty shales, clayey marls, and concretionary limestones are true water-formed strata, and we perceive in their numerous alternations and varying compositions the recurrent sediments of open and free-flowing seas.

But if the mere lithological composition of its rocks can thus throw light on the geographigal conditions of the Old Red Sandstone period, much more are we aided by a consideration of its fossils—the plant-life and animal-life that peopled the lands and waters. Wherever it has been examined, the *flora* appears to be of a lowly character—sea-weeds marsh or rush-like plants, clubmoss-like twigs, fronds of ferns, and less evidently, perhaps, drifted fragments of coniferous trees. We thus get a glimpse, as it were, of rocky weed-covered beaches, low marshy river-banks, of shady nooks and corners where fern and clubmoss luxuriate and of higher uplands fitted for the growth of coniferæ or pine-like trees. So far as known in Europe, the plants of the Old Red generally appear in detached and drifted fragments, and rarely in such abundance as to form a bituminous or coaly shale; but in Canada the thin seams of coal discovered by Dr Dawson would indicate not only a greater luxuriance, but even land areas on which they grew and died till the accumulated masses were sufficient to form successive layers of pure and crystalline coal. This is all that we learn of the dry land of the period from the vegetation that clothed it. We know nothing of its extent or configuration, nothing of its hills or valleys, of its lakes or rivers and are only left to infer from the nature and amount of the stratified sediments that the latter must have been both large and powerful. No true terrestrial creature—insect

probable, though requiring for its final demonstration a much more protracted and careful examination than the several phenomena have yet received from geologists.

reptile, bird, or mammal—has yet been detected in its strata ; and all that we know of its *fauna* is strictly aquatic, and in all likelihood marine.

This fauna of the waters differs, of course, like the existing fauna, in different seas ; but viewing the whole, and taking the entire range of the system through its lower, middle, and upper divisions, we have illustrations of the following zoological orders :—Corals, encrinites, and shells occur abundantly in the limestones of Devonshire, but similar organisms are altogether absent from the red sand-stones of Hereford and Scotland, and to a great extent also from the strata as developed in the north of Europe. Whether this has arisen from some peculiarity in the sea-bottom, or, as has been suggested, from the Scotch beds being chiefly of fresh-water origin, has not been satisfactorily determined ; but the fact stands undoubted that up to the present time no trace of a coral, an echinoderm (star-fish or encrinite), or a shell-fish has been detected in the Old Red Sandstone of Scotland. With what portion of the Scottish beds the Devonshire strata may have been contemporaneously deposited has not yet been determined ; but clearly it was not with the lower flagstones and bouldery conglomerates of Perth and Forfar. The Devonian corals and encrinites imply waters of genial temperature ; the bouldery conglomerates the reverse : and in all likelihood the two, though classed under the same system, were chronologically separated by ages.* But while the coral-building zoophytes,

* This is not, perhaps, the place to enter into the question of co-ordination ; but we cannot refrain from repeating our conviction, expressed in 1856, that the term "Devonian" can never be legitimately substituted for that of "Old Red Sandstone." We have examined the strata of Devonshire from north to south and from east to west, and instead of finding the equivalents of the Scottish Old Red we discovered in the Northern division one set of rocks that should be ranked with the lowermost Carboniferous, and in the Southern another that was perhaps contemporaneous with portions of the middle and upper Old Red. At all events, the

the encrinites, star-fish, and shell-fish seem thus to have had a partial distribution in the waters of the period, the annelids or worms, the crustacea, and the fishes abounded throughout, and this in numerous and varied specific aspects. Trails and burrows occur in every division of the system, from mere thread-like windings on the surfaces of the strata to burrows in the sandstones as thick as a man's arm; and crustacea (trilobites and eurypterites) throng especially the lower division in strange and often gigantic forms. Indeed the huge lobster-like forms of these eurypterites—*eurypterus, pterygotus,* and *stylonurus*—with their long segmented bodies void of appendages, and their broad carapaces, like the king-crab's, with limbs and jaws beneath, are characteristic features of the Old Red fauna. Ranging from three to six feet and upwards in length, with their toothed prehensile claws and oar-like swimming feet, no crustacean form has since equalled them in size, though few, perhaps, are more rudi-mentary in their structure. Like most articulated animals, these crustacea seem to have been readily dismembered by decay, hence their limbs and segments are frequently detached and scattered; and yet so wonderful has the preservative process been, even in the midst of dismemberment and de-cay, that their egg-packets, or masses of spawn (known as *Parka decipiens,* from Parkhill in Fife, where first detected) are common throughout the lower flagstones. How imper-ishable the record, could we only lay it bare, that nature keeps of her bygone aspects and operations !

rocks of Devonshire as a whole do not represent the Old Red Sandstone of Scotland, of Northern Europe, and North America as a whole ; and hence the inappropriateness of *Devonian* as a substitute for the earlier and more descriptive term *Old Red Sandstone.* The designation may yet be found to be an appropriate one for a set of formations that apparently lie between the true Old Red and the Carboniferous proper ; but to em-ploy it as synonymous with what was originally understood as the Old Red Sandstone system is, in our opinion, an error and misapplication.

Strange and gigantic, however, as are these early crustaceans, they are comparatively recent discoveries and but little known, and it is chiefly through its fossil fishes, their
numbers, variety, and beauty of preservation, that the system has become the subject of popular interest and investigation. As might be anticipated, these fishes differ considerably in the different portions of the system—those of the
lower being chiefly small or moderate sized, covered with
minute enamelled scales, and very generally armed with fin-
spines; those of the middle portion, again, being fewer in
number but larger in size, and protected by broad sculptured
scales or plates; while those of the upper zone, though still
covered with enamelled scales, assume more the character
of ordinary fishes, both in their size and configuration.
Throughout the whole, the bony enamelled scales and
plates (the exo-skeleton of anatomists) is the prevailing
feature, and all without exception are characterised by the
heterocercal or unequally-lobed tail—the upper lobe extending in a bold and prolonged sweep, as in the existing sharks
and dog-fishes. In Britain the great repositories of Old
Red fishes have hitherto been the lower shales of Forfarshire,
the lower and middle flagstones of Caithness and Cromarty,
the middle sandstones of Moray and Banff, and the upper
yellow sandstones of Dura Den in Fifeshire. In some of
these localities they are crowded together in shoals, with every
fin and scale in place as if overtaken and entombed by some
sudden catastrophe; and we have seen a slab about the size
of an ordinary writing-table, raised in Dura Den, with upwards of fifty individuals upon it, belonging to five separate
genera, and varying in length from ten to thirty inches.[*]

[*] At the instance of the British Association, and under the superin-
tendence of the late Dr Anderson, of Newburgh, and the Author, this
and numerous other slabs of nearly equal richness were raised from
Dura Den in 1860 and 1861; and could they have been rendered readily
portable, slabs of double these dimensions, and with treble the number

It would be out of place in a sketch of this nature to
enter into technical details, but it may be mentioned as of
some value, and not difficult of comprehension, that the
fishes of the lower zone, with fin-spines and minute lozenge
shaped or but slightly-rounded scales, are known by such
names as *acanthodes* (spiny), *cheiracanthus* (fin - spine)
diplacanthus (double-spine), *isnacanthus* (slender-spine)
parexus (ladder-spine), and so forth, in allusion to the
character of their spines ; that those of the same zone
having the head enclosed in a bony shield or series of
plates, are named *cephalaspis* (buckler-head) and *pteraspis*
(buckler-wing); that those of the middle zone having
their bodies enclosed in a bony case, somewhat like the
living trunk-fish, are known as *coccosteus* (berry-bone)
and *pterichthys* (wing-fish) ; while those of the same zone
with ordinary scales and fins are spoken of as *osteolepis*
(bony-scale), *dipterus* (double-fin), and *diplopterus* (twin-fin)
and that those of the upper zone, with their variously sculp
tured scales and head-plates, are known as *holoptychius* (all
wrinkle), *glyptolepis* (carved-scale), *glyptolæmus* (carved
throat), and other such names, having allusion to some well
marked and obvious distinction.   There is nothing very
puzzling in the names once their meaning has been explained
and the objects to which they refer have been examined
Indeed the local names for the living fishes of our own
coasts are often as puzzling and far less euphonious.   Go to
Cornwall and you hear one name, cross to Lincoln and you
have another; proceed to Fife and you hear a third, or north
ward to Wick and you have a fourth—all requiring ex
planation, and, till explained, as unintelligible as the much
vituperated technicalities of the palæontologist.

of specimens, could have been easily obtained.   The genera were chiefly
Holoptychius, Glyptolepis, Phaneropleuron, Glyptolæmus, Glyptopomus
and *Pterichthys.*

Beyond fishes, we know for certain of no higher life during the period of the Old Red Sandstone.   It is true that
remains of reptiles and reptilian footprints have been found
in the sandstones of Lossiemouth and Cummingstone in
Morayshire, but there are doubts about the age of these
strata—whether they be truly uppermost Old Red, or belong perhaps to the New Red or Triassic.   In this state of
uncertainty it may be generalised (provisionally, of course,
and having this doubtful instance fully in view) that the
flora of the Old Red period, scantily and obscurely developed, consists mainly of sea-weeds, marsh-plants, clubmosses, ferns, and coniferous-looking trees ; and that its
fauna, on the other hand, taking all the divisions of the
system as known in Europe and America, consists of corals,
encrinites, star-fishes, polyzoa, shell-fish, crustacea, and fishes.
We have thus no insects, no undoubted instance of reptiles,
no birds, no mammals.   No doubt the record is imperfect,
and it cannot for a moment be supposed that geologists in
the few scattered patches they have examined have detected
all, or nearly all of the Old Red Sandstone organisms.   Indeed, the existence of those already discovered necessarily
implies the presence of others on whom they preyed, or by
whom they were in turn preyed upon ; and the links we
have discovered in the chain of life, separated as they are,
prove the existence of the missing ones as clearly as if they
had been displayed before us.   Still, notwithstanding all
these facts and logical inferences, the flora and fauna of the
Old Red Sandstone curiously coincide in the main with all
that geology knows of the chronological development of life
on our globe, and we perceive in its discovered forms the
gradually-ascending steps in the great systemal scale of
vitality.

Such is a brief review of the Old Red Sandstone—a

period during which vertebrate life made its decided appearance on our planet, and during the continuance of which
several new distributions of sea and land were effected. We
say new distributions of sea and land, for there is no other
way of accounting for the differences that exist between its
lower, middle, and upper portions without supposing that
they were deposited in seas of different depths, and in seas
that derived their sediments from different directions. And
as these varying distribuions of sea and land necessarily
imply variations in climate and external conditions, we can
readily perceive how the plants and animals of the lower
portion differ from those of the middle, and these again
from those of the uppermost division. Nature is incessant
in her operations, and while the system of Waste and Reconstruction, described in our Sketch No. 2, endures, new
distribution of sea and land will be brought about in the
course of ages, varying conditions of climate will be effected,
and under the new conditions, some plants and animals
will shift their ground, some will flourish more luxuriantly,
and others again become altogether extirpated. But this is
not all: under these ever-varying conditions, and as time
rolls on, some forms of life seem to run their appointed
course and die out, and other and newer forms, in the
course of creation, seem to make their appearance. It is
thus that some forms of life are peculiar to the Old Red
Sandstone—that is, do not occur in earlier systems, and are
not found beyond the close of the period. Many forms of
coral, several genera of shell-fish, some trilobites, the gigantic crustaceans, pterygotus and stylonurus, the cephalaspis,
pteraspis, coccosteus, pterichthys, and other fishes, have
never been detected beyond the limits of the Old Red formation. They came in during the system, and died out
before its close; thus implying not only long lapses of
growth, and reproduction, and decay, but an onward march

in that creative process by which the world has been peopled by different and higher races during the advancing periods of its geological history. How wonderful this newer knowledge of life which geology imparts ! how marvellous the ever-ascending yet never-completed scheme of vitality it reveals ! To our forefathers the life of the present era was but a repetition of the life of former ages ; to us the life of the present is but a passing aspect, differing from the thousand aspects that went before, but inseparably bound up with them in one great scheme of ever-varying yet ever-widening development.

Such once more is the Old Red Sandstone, a system that owes its interest much more to its scientific than to its economic importance. Indeed, with the exception of building-stones used for local purposes, some indifferent lime-stones, and paving-flags, such as those of Caithness and Forfar, there are no rocks of any commercial value among its strata; and the only accidental minerals we are aware of are occasional poorish veins of galena, veins of baryta, salt springs like those of North America, and the pebbles of agate, carnelian, and the like (Scotch pebbles), obtained from the amygdaloidal trap-rocks that traverse the system. Its chief interest centres round its fossil fishes and crustacea, subjects rendered popular now more than twenty years ago by the writings of Hugh Miller and Agassiz, and still attracting attention by the newer forms that are year after year made known by the labours of younger geologists.* And surely what geologists are labouring to reveal, the man of ordinary intelligence may make some effort to comprehend and en-

* We allude in particular to the labours of Professor Pander among the Old Red fishes of Russia ; the numerous discoveries of new crustaceans and fishes in the flagstones of Forfarshire by Mr Powrie; and the long-continued observations of Dr Gordon among the sandstones of Moray and Ross-shire.

joy. It must indeed be a dull and incurious mind tha
cannot be induced to take an interest in the history of th
world he inhabits, and to trace in its formations the recor(
of operations which took place, and the nature of being
that lived and died, thousands of ages before the humai
race was created to become participators in the same evei
varying and ever-advancing scheme of vitality. Astronom
may be a loftier theme, but the loftiness of its topics onl
renders them the colder and more remote. Geology, on th
other hand, has ever an immediate and human interest
The Earth's Past is inseparably interwoven with her Pre
sent ; that which now lives is intimately associated in plai
and relationship with that which lies fossil in the rock
beneath us ; this plan has been steadily evolving durin{
untold ages ; man's own history is but part and parcel o
that plan ; and surely whatever tends to exalt our concep
tions of creation can never tend to weaken our reverenci
for the power, wisdom, and goodness by which it is directec
and sustained.

COAL AND COAL-FORMATIONS.

COAL, ITS ORIGIN AND FORMATION—MINERALISED VEGETATION—
RECENT PEAT-GROWTHS—TERTIARY LIGNITES—SECONDARY AND
PALÆOZOIC COALS—PRIMARY ANTHRACITES AND GRAPHITES—THE
COALS AS A MINERAL FAMILY—CONVERSION OF VEGETABLE SUB-
STANCES INTO COAL—ITS VARIOUS STAGES—PEAT, LIGNITE, COAL,
ANTHRACITE, AND GRAPHITE — CHARACTERISTICS OF THESE RE-
SPECTIVE STAGES—IMPORTANCE OF COAL TO CIVILISED COUNTRIES
—SPECIAL VALUE OF, TO GREAT BRITAIN.

THERE is no mineral in the crust of the earth more essential
to modern civilisation than coal, and there is, perhaps, no
geological technicality more frequently made use of than
" Coal-formation," and yet how few have a rational or intelligent conception of either ! Every man and woman in
the British Islands is less or more acquainted with the ordinary aspects of coal and its uses, and yet not one in a hundred, perhaps, could give the commonly received explanation
of its nature and origin. Most people are aware that coal
is obtained by mining in rocks known as the Coal-formation, and yet how few know anything of the nature of
these rocks, how they were aggregated, or by what means
coal was formed along with them ! It is true that men of
science have their differences about these things, as they
have about many other matters ; but these differences are
for the most part trivial, and do not affect the general belief
either as to the nature of coal or the processes by which it
was aggregated. It is to state these beliefs in a simple and
intelligible way that we attempt the present Sketch, and

our main object will be to exhibit the points upon which
geologists are generally agreed, rather than to distract the
non-scientific reader with the minutiæ upon which some of
them still continue to differ.

What is coal? is a question more satisfactorily answered
by a little roundabout explanation than by a direct reply.
To say that coal is altered and mineralised vegetable matter
is true; but the definition is too curt to be readily intelligible.   Every one knows something of peat and peat-mosses;
well, this *peat* is simply coal in its first stage of development.   Were the peat-moss submerged and covered over
by deposits of mud and clay and sand, it would in course
of time undergo important chemical changes, by which
part of its gaseous contents (oxygen, hydrogen, &c.) would
be discharged, and the mass reduced to a compact coaly
substance known as *lignite* or *brown-coal*.   Such brown-
coals are abundant in many countries (Germany, Austria,
New Zealand, &c.), and worked for economical purposes;
and were they subjected to still further changes they would,
in course of ages, become converted into shining *stony coals*
like those which are now raised so largely from the coal-
fields of Great Britain.   The truth is, coal occurs in the
earth's crust in every stage of development, from the peat-
mosses and swamp-growths still in process of accumulation
on the surface, down through the tertiary brown-coals to
the bituminous stone-coals of the secondary and primary
periods, and from these again down to the still older non-
bituminous *anthracites* and *graphites*.   All, in fact, have
had a similar origin.   They are mere vegetable masses that
have undergone different degrees of mineralisation — the
recent vegetable full of volatile matters, the lignites less
so, the bituminous coals giving off smoke and flame, the anthracites barely smoking, and the graphites masses of pure

debitumenised carbon. They are all coals, and belong to the same family—those in the younger formations still retaining much of their vegetable structure and full of volatile matter, while those in the older formations have seemingly lost all traces of structure, and have been all but deprived of their volatile constituents. But even where no structure is obvious to the naked eye, it can generally be rendered apparent by submitting thin transparent slices to the microscope. By this means the vegetable origin of the most compact and glistening coal is often revealed as clearly as the tissues in living plants, and thus the observer is enabled to determine not only the organic nature of the mass, but the botanical peculiarities of the order concerned in its formation.

Since coal is thus merely altered and mineralised vegetable matter, and since vegetation must have flourished more or less during every period of the earth's history, there must be coals of some kind or other occurring in every geological formation. It may appear more abundantly and more availably in one formation than in another; still, believing in the uniformity of nature's operations, we must be prepared to admit its presence in every stratified system, and not to regard it, as was at one time done, as a product peculiar to the Carboniferous era. Arranging the formations in chronological order, we have their coals, or rather the coal family, associated with them in something like the following conditions :—

Quaternary, ... *Peats.*
Tertiary, ... *Lignites.*
Cretaceous, ... *Lignites and Coals.*
Oolitic, ... *Coals.*
New Red Sandstone, ... *Coals.*
Carboniferous, ... *Coals and Anthracites.*
Old Red Sandstone, ... *Coals and Anthracites.*
Silurian, ... *Anthracites.*
Cambrian, ... *Anthracites and Graphites.*
Laurentian, ... *Graphites.*

K

It is true that in the older formations we have but a very scanty exhibition of coaly substances, and it is equally true that hitherto the most extensive developments have been found in strata of Carboniferous age ; but it is nevertheless the fact that coal-fields of great value occur in the oolitic and cretaceous rocks, and that brown-coals are common in almost every tertiary district. It may render the subject more intelligible and attractive if we take the formations seriatim — beginning with the recent and apparent, and working down through the older and more obscure.

The coals of the present day are the peat-mosses, the swamp-growths, and the vegetable drifts borne down by rivers and deposited in their estuaries.   We have no means of ascertaining the extent or thickness of vegetable drifts, though some, like the " Rafts " of the Mississippi, are of considerable thickness and extent; but we know that large areas in all the temperate and colder latitudes are occupied by peat-mosses and swamp-growths—the lake region of North America, Canada, the Southern States, Siberia, Northern Europe, Denmark, Holland, and our own islands.*   These are often of great thickness, and date from the growth of the current year to the very dawn of the Quaternary epoch ; loose and turfy above, firm and peaty a few feet down, and at greater depths black and dense as some varieties of lignite.   Indeed, we have seen varieties of Dutch peat taken at 30 feet deep indistinguishable from some lignites ; and

* We have no reliable statistics of the extent and thickness of peat-mosses either in Europe, Northern Asia, or North America ; but in the recently published Report on the Geology of Canada by Sir William Logan, a number of details are given, from which we learn that upwards of 300 square miles of that country are occupied by peat-mosses varying from 3 to 30 feet in thickness.   If such be an approximation to the amount of peaty surface in the surveyed portion of Canada, the amount in the whole of British North America, Northern Europe, and Northern Asia, must be something enormous.

all that seems necessary to convert them into true brown-coals are the cover and pressure of superincumbent strata, and time sufficient to effect those further chemical changes to which lignites and brown-coals have been generally subjected. We see, therefore, in the compressed vegetable matter we call *peat*, and which has been formed by the growth and decay of certain plants * during many centuries, the first stages of *coal*, and when we come to consider the older formations, we shall find that many of their coal-seams have had a similar origin. And just as this peat is sometimes earthy and mingled with stony matter that has been washed into the swamps and hollows by rains and rivers, so we may expect some of the old coals to contain similar impurities, and to be less valuable as fuel.

The next and older series of coals embraces the *lignites* or *wood-coals*, the *brown-coals* and *board-coals* of the Tertiary strata. As these names imply, their woody or vegetable texture is still apparent, and they are generally of a brown or earthy hue, compared with the black and glistening lustre of the coals of the older formations. Alternating with clays, marls, sands, and gravels, they have evidently been formed partly in fresh-water lakes and swamps, and partly in areas that have been submerged and covered over by marine deposits. In some instances they are earthy, and composed of the drifted trunks and branches of trees; and in others the submerged and fallen forest-growth can be traced as clearly as it can be in some of the shallower peat-beds of Scotland. Most of these lignites, whether as once worked at Bovey in Devonshire, or as still worked in Germany, Prussia, Austria, New Zealand, and other countries, may be described as coal in its second stage of consolidation and mineralisation. In the mine they are

* See Sketch entitled " Recent Formations."

soft, full of water, and easily cut; and when brought to
the surface, dry and break up, and soon crumble down
under the influence of the weather.   They are also less
regular in their bedding than the older coals—thickening
and thinning capriciously; but in some instances their
bedding is regular and continuous over considerable areas,
and their quality is so much improved that they are scarcely
distinguishable from ordinary coal.   One remarkable instance of this kind, the Zsil valley in Transylvania, was
visited in 1862 by Professor Ansted, who found not lignite, but coal differing little from some varieties of English
coal, lying in regular beds of great thickness, and alternating with shales, ironstones, and grits.   Of course, all the
vegetable accumulations of the Tertiary system are not
precisely of the same age, nor have they been deposited
under the same conditions, and thus we may expect to find
differences among them, just as among the coals of the older
formations.   And hence it happens that some of these
lignites are scarcely fit for pottery or brick-kiln purposes,
while others (certain compact and lustrous varieties) are
advantageously used for locomotive engines and for metallurgical operations.

Although seams of lignite are occasionally found in the
Cretaceous and Oolitic systems, yet, generally speaking, the
Secondary strata—the Chalks, Oolites, and New Red Sandstones—are characterised by the presence of true coals.
The seams may not be continuous over extensive areas—
that is, may thicken and thin somewhat capriciously—but
still mineralisation of the mass is complete, and we are
presented with bituminous coals of varying commercial
value.   Such Secondary coal-fields occur at Brora and
Whitby in Britain; at Fünfkirchen and Oravicza in Austria; at Burdwan, Nerbudda, and other districts in In-

dia; in Burma, Borneo, and Labuan; in New Zealand; Natal in South Africa; Vancouver Island, and British Columbia; at Richmond in Virginia; and in all likelihood in other districts that have not yet been sufficiently sur-veyed. These coals, so far as they have been geologically examined, have been accumulated precisely like the peat-mosses, swamp-growths, and vegetable-drifts of the present day. Some have evidently grown *in situ*, and accumulated in great thickness and purity for ages; in others the growth has been interrupted by overflowings of the water, and earthy layers are not unfrequent in the mass; while in others, again, the mass is so irregular in thickness and composition, that it at once recalls the idea of drift and heterogeneous deposition. Whatever the thickness or composition, they are true bituminous coals, thus disprov-ing the belief which was generally entertained some twenty or five-and-twenty years ago, that all true coal was a product of one geological epoch only, and necessarily belonged to the Carboniferous formation. On this point we can offer nothing more convincing than the following extract from Sir Charles Lyell's description of the Richmond coal-field in Virginia:—"These Virginian coal-measures are com-posed of grits, sandstones, and shales, exactly resembling those of older or primary date in America and Europe, and they rival, or even surpass, the latter in the richness and thickness of the coal-seams. One of these—the main seam—is in some places from 30 to 40 feet thick, com-posed of pure bituminous coal. On descending a shaft, 800 feet deep, in the Blackheath mines in Chesterfield county, I found myself in a chamber more than 40 feet high, caused by the removal of the coal. Timber props, of great strength, supported the roof; but they were seen to bend under the incumbent weights. The coal is like the finest kinds shipped at Newcastle, and when analysed yields the same proportions of carbon and hydrogen—a fact worthy of notice when we consider that this fuel has been derived from an assemblage of plants very distinct specifically, and in part generically, from those which have contributed to the formation of the ancient or palæozoic coal."

"It is true, however," says one of the most experienced and practical of British geologists (Professor Ansted), "that the great coal-fields of England, of Belgium, of Spain, of France, and of North America, besides those of Bohemia, Moravia, and the Rhine, of Russia and China, and probably of Australia, belong to the oldest or palæozoic rocks, and that for some reason that may perhaps be better understood at a future time than it now is, these deposits are more regular, more uniform over large areas, and in that sense more to be depended upon, than those of newer date." * In other words, they belong to the Carboniferous system, that great series of limestones, sandstones, shales, ironstones, and coals, which has hitherto yielded the main supplies of mineral fuel, and to which Britain owes so much of her mechanical superiority and commercial greatness. As this system will form the subject of a separate Sketch, we need only here observe that its coals occur in many seams, of every thickness, from a few inches to forty feet ; of all degrees of purity, from earthy masses that

---

* There can be no doubt that the difference here alluded to has arisen, partly from the peculiar distribution of sea and land during the Carboniferous era, which permitted over extensive areas a moist, genial, and equable climate, and partly from the peculiar character of the vegetation of the period, which seems to have been at once of rapid growth and of a kind eminently fitted for preservation. Physical conditions like a moist, genial, and equable climate may recur in the course of nature, but the Life of each geological system is peculiar, and vanishes with the period to which it belongs. The Carboniferous flora disappeared with its epoch ; and no flora equally fitted for the formation of coal has since recurred or may ever again recur in the progressional course of creation.

can scarcely be ignited, to clear bituminous seams that burn leaving scarcely a trace of ashes; and fitted for every economical purpose—household fuel, gas-making, oil-distillation, steam-raising, smelting, and metal-working. Like other coals, the thickest, the purest, and most continuous seams have evidently grown and accumulated *in situ;* those imbedding stony and earthy layers have been interrupted in their accumulation; and others, again, less regular and continuous in thickness and mingled more with extraneous impurities, have apparently been formed of drift and water-logged vegetation. In lakes, in estuaries, and along great shallow sea-reaches, the flora of the Carboniferous era flourished for ages, the land now sinking, now rising, but on the whole subsiding, to receive the vast thickness of sediments which compose the system. Read in the light of what is now taking place at the present day, there is nothing abnormal or preternatural in the Coal-formation, and we behold in its various coals merely the peat-growths, swamp-growths, jungle-growths, and vegetable-drifts of the period, compressed and mineralised during the lapse of ages. The caking-coals, splint-coals, cannel-coals, and anthracites or stone-coals of the miner are merely different expressions of this mineralisation or metamorphism—different conditions of deposit, as rapid covering up, exposure to decay, nature of vegetation, and compactness of overlying strata, all affecting the ultimate quality of the coal. If much earthy matter has mingled with the vegetable mass during its aggregation, the coal will be stony and impure; if the vegetable mass has been rapidly covered up by retentive muds and clays (now converted into shales and fire-clays), the coal will likely be soft and highly bituminous; if the superincumbent stratum be open and porous, so as to admit the escape of volatile matters, the coal will in all likelihood be hard, dry, and less bituminous; and if the vegetable mass has undergone extreme chemical change, or has, as a coal, been subjected to the heat of igneous rocks, it will less or more be deprived of its gaseous elements and converted into an anthracite.   And thus, and thus only, can the great variety of coals occurring in the palæozoic coal-fields of Europe and North America be satisfactorily accounted for.

Beyond the Carboniferous system coals become rare and comparatively unimportant.   It is true that in some districts we cannot fix any very sharp line of demarcation between the Coal-formation and the Old Red Sandstone, but generally speaking the two systems are sufficiently distinct, and it is curious that up to the present time no coal-seams of any thickness have been detected in the latter.   Indeed, with the exception of some insignificant bands described by Principal Dawson as occurring at Gaspé in Canada, the Old Red Sandstone is altogether barren of coal, though vegetable fragments are scattered in some abundance throughout its shales and flagstones.   In the Silurian and more highly metamorphosed Cambrian and Laurentian strata we have thin bands and irregular patches of anthracite and graphite ; but though these are generally ranked with the coal family, their vegetable structure has been so obliterated that we cannot say whether they have been formed from terrestrial or marine vegetation, or indeed whether graphite is always certainly of organic origin.

Here, then, we perceive that *Coals*, or minerals of the *Coal Family*, occur in all formations, from the accumulations now going forward on the earth's surface down through every stratified system, whether belonging to tertiary, secondary, or primary epochs.   From peat we pass to lignite, from lignite to true coal, from coal to anthracite, and from anthracite to graphite. All are but compressed and chemically altered masses of vegetation, the slow fermentation or distillation of which results in the gradual expulsion of the gaseous or volatile portions, and in the retention of the carbonaceous or coaly residue.* The following tabulation exhibits, proximately, this gradation of chemical change by which wood is converted into peat, peat into lignite, lignite into coal, coal into anthracite, and anthracite into graphite:

(At 212°.)	Carbon.	Hydrogen.	Oxygen.	Nitrogen.	Inorganic Ash.
Wood............	48—54	6—10	35—45	...	...
Peat..............	56—66	5— 9	18—33	2—4	1— 6
Lignite...........	56—70	3— 7	13—27	1—0	1—13
Coal..............	70—92	2— 6	1— 8	0—2	3—14
Anthracite......	74—94	1— 4	0— 3	trace	1— 7
Graphite	80—98	...	...	...	1— 7

Here it will be observed that the gaseous substances, hydrogen and oxygen, so abundant in recent wood and peat, gradually diminish as the mass becomes more and more mineralised, till at length they disappear and leave in consequence a gradually-increasing residue of carbon in the true coals, anthracites, and graphites. Like all mixed rocks, however, coal presents itself in many varieties. We cannot conceive of vegetable matter (whether drifted or grown

* According to M. Fremy, the following are the degrees of alteration of woody tissue : 1. *Turf and Peat.*—Characterised by the presence of ulmic acid, and also by the woody fibres or the cellules of the medullary rays, which may be purified or extracted in notable quantities by means of nitric acid or hydrochlorites, in which they are insoluble. 2. *Fossil Wood* or *Woody Lignite.*—This, like the preceding, is partially soluble in alkalies, but its alteration is more advanced, for it is nearly wholly dissolved by nitric acid and hydrochlorites. 3. *Compact* or *Perfect Lignite.*—This substance is characterised by its complete solubility in hydrochlorites and in nitric acid. Alkaline solutions do not in general act on perfect lignites. Reagents in this variety show a passage of the organic matter into coal. 4. *Coal.*—Insoluble in alkaline solutions and hydrochlorites. 5. *Anthracite.*—An approximation to graphite ; resists the reagents which act on the above-mentioned combustibles, and is only acted on by nitric acid with extreme slowness.

in situ) being associated with sedimentary strata without its being mingled more or less with the earthy impurities of these sediments. These impurities, according to their amount, must necessarily confer on different coals different structures, different aspects, and different qualities. Besides, varieties will also arise from the conditions of the vegetable mass, itself, according as it may have been imbedded while fresh or been long exposed to atmospheric decay, according as it may have been suddenly covered up or long exposed to maceration and comminution in water, and notably also according to the nature of the plants composing the mass. These varieties, according to their structure, texture, and qualities, are generally known as *caking-coal*, which is soft and tender in the mass, like that of Newcastle, and swells and cakes together in burning; *splint* or *slate coal*, which is hard and slaty in texture, like most Scotch coals, and burns free and open; *cannel* or *parrot coal*, which is compact and jet-like in texture, spirts and crackles when thrown suddenly on the fire, but when ignited burns with a clear candle-like flame, and from its composition is chiefly used in gas-manufacture; and *coarse*, *foliated*, or *cubic coal*, which is more or less soft, breaks up into large square blocks, and contains in general a large percentage of earthy impurities. Between these varieties there is, of course, every gradation—coals so pure as to leave only one or two per cent of ash, others so mixed as to yield from ten to thirty per cent, and many so impure as to be unfit for fuel, and so to pass into *shales* more or less bituminous.[*]

---

[*] As bituminous *shales* are now so extensively mined for the distillation of paraffin, it may be of use to advert to some distinctions that subsist between them and the *coals* properly so called. A *coal*, though often containing a considerable amount of earthy impurity, consists chiefly of vegetable matter, or, in other words, carbon is its prevailing ingredient. Where the earthy or mineral ingredient greatly exceeds the organic, it becomes unfitted for combustion, and is regarded merely as a carbonaceous *stone*,

Besides these varieties, founded chiefly on mineral characters, it is also customary to distinguish coals according to the purposes for which they seem best suited, or to which they are most frequently applied; hence we hear of *household coals, furnace coals, smithy coals, steam coals, gas coals, oil coals,* and similar distinctions.

We have thus occurring in the crust of the earth not only a great variety of coaly substances, but also coals of different aspects and qualities occurring in the same geological formation. The causes of these differences are, in general, sufficiently obvious:—age, and the amount of chemical change to which they have been subjected ; the amount of earthy impurities commingled with them during their aggregation and deposition ; the nature of the plants composing the bulk of the mass; the amount of decay which the vegetable mass had undergone before it was finally covered by other strata; and the porous or retentive nature of the strata between which it is imbedded. All these and other causes have tended to create the differences that now exist among the different members of the Coal Family—the

of which clay, sand, and the like form the main proportion. The term *shale,* on the other hand, refers to structure rather than to composition, and is something that splits up or *peels off* in thin layers or laminæ. Most consolidated muds are characterised by this quality of splitting or breaking up in thin leafy layers parallel to their bedding; hence shales may be regarded as consolidated muds, and may be distinguished as calcareous, arenaceous, or bituminous according to their predominating ingredient. Bituminous shales, therefore, have been mere vegetable muds—their richness, like those of the coals, depending upon the amount of organic matter and the conditions under which it was preserved. Some shales may be as bituminous as some poor varieties of coal, but this does not entitle them to be ranked as coals, any more than an excess of earthy matter in a hard stony coal would entitle it to be called a shale. The terms refer to structure rather than to composition ; and though it is true that the shaly or leafy structure is almost invariably characteristic of the earthier ingredient, yet it must ever be borne in mind that both shales and coals are *mixed rocks,* and that not unfrequently the one may pass into the other by insensible gradations.

graphites and anthracites burning like charcoal, without smoke or flame; the ordinary bituminous coals burning with varying degrees of smoke and flame; the lignites burning with stifling odour, and expelling much watery vapour and smoke; and the peats scarcely combustible till dried in the sun or by hydraulic pressure, and then burning with little flame but with much smoke and their own peculiar odour. But whatever their peculiarities in these respects, they are all highly important substances, and stand along with iron as the most valuable that human industry obtains from the crust of the earth.   Indeed, the coals are altogether indispensable to modern civilisation, the peculiar mechanical phases of which are mainly of their own creating.   So long as man depends upon the forests for his fuel, his mastery over the metals is limited, and his mechanical appliances restricted.   But when he has once learned the uses of coal, and can obtain it in fair supplies, his metal-working powers expand, and his forges, factories, steam-engines, steam-ships, gas-works, railroads, and electric telegraphs become the necessary developments of this new acquirement.   Once acquainted with these and similar appliances, man takes his stand on a higher platform, gains new ascendancy over the forces of nature, and overcomes in a great measure the obstacles which time and space oppose to his operations.

Where and at what time man first began to employ coal as a fuel is unknown.   The Chinese and Japanese have evidently been long acquainted with its uses, but their chronology is uncertain.   The Hindoos, Egyptians, and other Oriental nations never seem to have searched for any variety of mineral coal, but laboriously prepared wood-charcoal for their metallurgical processes.   The Greeks and Romans were acquainted with its properties, though they appear to have seldom employed it, and this on the most limited scale.*   In our own country some ancient crop-workings, with stone hammers and hatchets still remaining, date back perhaps to the Roman invasion, and "coal" is mentioned in Saxon records of the ninth century; but it was not till the twelfth and thirteenth centuries that the value of the substance began to be fairly recognised.   And in connection with these facts it is a circumstance worth noting, that no savage race has ever yet been discovered that seemed to be aware of its nature and uses.   It may crop out along the ravines and sea-cliffs, as it does in North America, in Eastern Africa, in Farther India, in Australia and New Zealand, but the savage never bends to dig while the twigs and branches around him can be broken.   The use of certain minerals and metals are, in truth, as satisfactory tests of man's progress in civilisation as the cultivation of certain plants or the domestication of certain animals.   The possession of the one may largely increase his comforts; a knowledge of the other invests him with new and higher powers.

As a nation we cannot exalt too highly the importance of our coals and coal-fields.   Our mechanical, manufacturing, and commercial greatness is intimately bound up with their existence; and whatever tends to disseminate a knowledge of their nature, to develop their resources, or economise their products, is worthy of our encouragement and attention.   Commercially, we may have no immediate interest

---

* It is thus referred to by Theophrastus, a Greek author, who wrote about 240 years B.C. :—"Those fossile substances that are called Coals, and are broken for use, are earthy; they kindle, however, and burn like wood coals.   These are found in Liguria, where there also is amber, and in Elis in the way to Olympias over the mountains.   These are used by the smiths."— *Sir John Hill's Translation*, 1774.   The Ligurian coal would appear, from its connection with amber, to have been lignitic; the Olympian coal, now being worked by a modern company, is bituminous and of older date.

in the substances called *coals,* but indirectly every man is
less or more indebted to their applications; and it must be
a dull or indifferent mind that cannot be induced to take
some interest in products to which his country owes so
much of her power and greatness, and himself so many of
the comforts and amenities he is daily enjoying.

THE OLD COAL-MEASURES.

THE reader who has perused the preceding sketch, will have
seen that coal is a product of every geological epoch, from
the peat now accumulating on the earth's surface down
through the lignites of the tertiary, the true coals of the
secondary, and the harder coals and anthracites of the primary periods. But though thus occurring in all stages of
the earth's history, it is in the so-called " Carboniferous System " that it appears in numerous seams, in many varieties,
and in great thickness and continuity over extensive areas.
It is from this old system that Britain, France, Belgium,
Russia, China, Australia, and the United States of America obtain their main supplies; hence the familiar terms
" Coal-Formation " and " Coal-Measures," as if it were the
only series of coal-yielding strata in the crust of the globe.
In Britain it generally rests on a series of reddish sandstones, and is in turn overlaid by another series of red
sandstones ; the former being naturally designated the

" Old Red," and the latter the " New Red," by the systematic geologist.   It thus holds a sort of middle place in
chronological classification, being younger than the Cambrian, Silurian, and Old Red systems, and older by far
than the Chalks, Oolites, and New Red Sandstones.   Its
position is well defined, and may be seen at a glance in the
following sequential arrangement :—

CAINOZOIC . (*Recent*).	{ Quaternary or Recent. { Tertiary.
MESOZOIC (*Middle*).	{ Cretaceous or Chalk. { Oolitic or Jurassic. { Triassic—(Upper New Red).
PALÆOZOIC (*Ancient*).	{ Permian—(Lower New Red). { CARBONIFEROUS—THE OLD COAL-MEASURES. { Old Red Sandstone and Devonian. { Silurian.
EOZOIC (*Dawn.*)	{ Cambrian. { Laurentian.

It is to these palæozoic or ancient coal-measures, in contradistinction to all others, that we direct the present
sketch, dwelling more especially on their geological aspects,
and only incidentally alluding to their industrial applications and importance.   We say " incidentally alluding ; "
for their building-stones, fire-clays, alum-shales, limestones,
ironstones, and coals—the labour and skill expended in
mining them, the innumerable uses to which they are applied, and their bearings on the industrial and social conditions of a people—are subjects which of themselves would
require the consideration of half-a-dozen sketches.

Perhaps the most intelligible way of treating any geological system, is to consider it *first* as a Rock-formation,
second as a Life-period, and *third* as an Economic repository.   In this way we get an insight into the nature of the
strata of which it is composed, and the agencies concerned

in their aggregation; into the character of its fossils, which thus throw light on the geographical conditions of the period; and, finally, into its industrial value, as bearing on the wants and progress of civilisation. Adopting this method, we find the Carboniferous system composed in the main of sandstones, shales, fire-clays, ironstones, limestones, and coals, all many times alternating with each other, and in some districts attaining to a thickness of 12,000 or 14,000 feet. Of course, during the deposition of such a vast thickness of strata, and which necessarily implies the lapse of long ages, there must have been frequent changes in the relative levels of sea and land; and hence some of these sediments were laid down in deep and others in shallow water, while the shallower beds were once more sunk to greater depths, and overlaid by newer sediments. In this way the Carboniferous system consists, in most regions, of several series of strata, and in the British Islands these are generally arranged and named as follows :—

1. Upper or true coal-measures.
2. Millstone grit or sandstone series.
3. Carboniferous or mountain limestone.
4. Lower coal-measures or carboniferous shales.

Although these several series have evidently been deposited in waters of various depths and under somewhat different geographical conditions—the lower being apparently more *estuarine*, the mountain limestone being more *marine*, the millstone grit more *littoral*, and the upper more *terrestrial*—still there is a great family resemblance, so to speak, between them, and, with the exception of the coal-seams, they are all strictly sedimentary, and bear in their structure and texture abundant evidence of the aqueous agencies concerned in their formation. In the sandstones and grits—often ripple-marked, rain-pitted, and worm-burrowed—we

trace the sands of open and exposed shores ;* in the shales
and fire-clays and ironstones, the muddy deposits of deeper
waters ; and in the limestones, which also vary much in
composition and character, the shell-beds, the coral-growths,
and zoophyte drifts, both of the brackish estuary and of the
outer ocean.   Of course, among sediments so varied we
may expect to find every degree of admixture—sandstones
pure, quartzose, and compact; sandstones flaggy, laminated,
and clayey ; and sandstones calcareous and coaly.   Shales
so purely argillaceous as to be termed fire-clays, shales calcareous, shales bituminous, and shales so sandy as to pass
into flaggy sandstones.   So also it is with the limestones ;
some so pure as to contain scarcely a trace of earthy matters,
and others so mingled with earthy impurities as to be altogether unfit for economical purposes.   The ironstones, too,
which were merely the ferruginous muds of the Carboniferous sea (chemically aggregated by the union of the carbonic
acid given off by decaying vegetation, and the iron held in
solution in the waters), appear as " clay-bands " or clay-
carbonates, as " black-bands " mingled more or less with
coaly matter, or as stony impregnations too poor to be
worked to advantage.

Respecting the coals, which we separate from the strictly
sedimentary beds, there are also many varieties both as to
composition and structural peculiarity.   Where the vegetable mass has evidently accumulated on the spot, as peat-
moss, swamp-growth, and forest-growth, the seam is generally

---

* Several of the thick-bedded sandstones of the British coal-fields have
evidently arisen, in the first instance, from Æolian or wind-blown sands,
like those that form the " links " and " dunes " of the present day.
Their whole internal arrangement points to this mode of aggregation,
though they have, of course, been subsequently submerged and planed
down on their upper surfaces by the action of water. Illustrative examples may be seen along the eastern shores of Fife, and especially
between Crail and St Andrews.

pure, and spread over a considerable area, with great regularity as to thickness and quality. Surprise has been frequently expressed at the uniform thickness which many coal-seams maintain over extensive areas. Growth of the vegetable mass *in situ* is no doubt the main cause, but we must not lose sight of the fact that, when in a semi-plastic state of bituminisation, the pressure from above would have a tendency to spread out the seam, and insensibly equalise its thickness. On the other hand, where interruption to this growth has taken place either from periodical inundations or otherwise, the seam contains layers of earthy matter, and is more or less impregnated throughout with such impurities. Again, where the seam has arisen from drifted vegetation, it is still less regular in thickness, and often so earthy and impure as to pass into a bituminous or coaly shale. These bituminous or coaly shales, now coming so largely into use in the Scottish coal-fields for the distillation of paraffin and paraffin oil, are indeed of very various origin and composition. Some of the richer sorts are merely compressed and mineralised vegetable muds that have arisen from long maceration and decay; others, during this long decay in shallow water, have got so largely mingled with the remains of minute crustaceans (cyprides, &c.), as to be partly of animal origin;[*] and some again of the poorer sorts are little else than thick clayey silts, irregularly intermingled with vegetable and animal debris. In this way the various purities of coal can be readily accounted for, while difference of mineral aspect and quality may have arisen partly from the nature of the vegetation, partly from

[*] One of the most remarkable we have examined is the " Grey Shale " of West Calder, raised for the extensive paraffin works of Mr Young, and which derives its name from the colour imparted to it by the calcareous cases of these minute organisms. In some places the seam, which is upwards of two feet thick, is literally a mass of these remains.

the rapidity with which it was entombed, partly from the porous or retentive character of the imbedding strata, and partly also from the degree of mineralisation the respective seams have undergone. Hence the soft *caking coals*, which fuse together in burning; the hard, slaty, *splint coals*, which burn dry and open; the coarse *cubic coals*, which also burn open and leave much earthy ashes; the compact, lustrous, *cannel coals*, used chiefly in the making of gas, and other varieties well known in one or other of our British coal-fields. There is no great difficulty, we repeat, in account-ing for the varieties of palæozoic coals, if we only make allowance for difference in the nature of the vegetation, in the modes of its accumulation, the length of time it was exposed to maceration and decay, the retentive character of the covering stratum, and the intensity of mineralisation which these different conditions would induce. Of course, all this implies long ages of growth and decay, repeated emergence and submergence of the land, but in the main a gradual subsidence to permit that vast accumulation of sedi-ments — sandstones, shales, ironstones, and limestones, to the thickness of many thousand feet—which constitute the bulk of the Carboniferous system.

Besides the strictly sedimentary strata, the coal-measures are also in some districts largely made up of igneous pro-ducts, which intermingle with them as masses of basalt and greenstone, beds of trap-tuff, and other vulcanic discharges. Of course, these discharges must have taken place in or near the seas of deposit,—now as overflows of lava, now as showers of ashes, and again as the mingled products of vol-canic eruption. Just as insular and submarine volcanoes are at the present day mingling their eruptive matters with the sediments of the adjacent seas, so in the old coal period similar agencies were at work, and the results are now the interstratified greenstones and trap-tuffs, the bent and fractured strata, and the filling-up of the rents and fissures with igneous rock-matter. We say "interstratified greenstones," for some observers speak of "intrusive greenstones," as if, during the vulcanic paroxysms, such igneous rocks had been forced for miles in every direction between the separated strata! There are, no doubt, intrusive masses among the strata of every formation, but these are generally limited in extent, irregular in form, and bake or harden alike the immediately underlying and overlying beds. The interstratified greenstones, on the other hand, affect only the strata on which they rest, and to regard such widespread lava-like sheets as *intrusive*, is simply an absurdity. Indeed, almost every feature of the vulcanism of the period has been perfectly preserved to us, and the imagination has little difficulty in recalling the broad bays and estuaries of the Carboniferous ocean, studded with their cones and craters of eruption—here ejecting showers of dust and ashes, there discharging floods of molten rock-matter, and ever and again the whole shaken and fractured by earthquake energy and convulsion. Part of this vulcanism was contemporaneous with the deposition of the sediments, as proved by the interstratified greenstones and trap-tuffs, but part also happened long subsequent to the solidification of the strata, as seen in the faults and dykes of injected matter that intersect the whole thickness of the system; but whether contemporaneous or long subsequent, it forms one of the most remarkable features in the coal-formation, engaging the closest study of the geologist, and exercising all the ingenuity of the miner and engineer. If the reader could picture to himself the district in which he resides fractured by earthquake convulsion—here a portion thrown many fathoms up, there another portion thrown many fathoms down, and the rents between filled with solidified lava—he would have before him precisely the appearance presented by the "upthrows and down-throws," the "faults and dykes" of many coal-fields, and especially of those of the Scottish Lowlands.

Composed, like other systems, partly of stratified sedi-ments, and partly of unstratified masses which were the volcanic products of the period, the Coal-measures present no great difficulty as a Rock-Formation, and few of its strata have undergone much metamorphism or internal change, unless where in contact with igneous eruptions. In its *stratified* rocks we perceive the obvious sediments of seas, lagoons, and estuaries, the relics of shell-beds and coral-reefs, the vegetable growths that accumulated for centuries in swampy morasses,* flourished in the virgin forests, or tangled rankly in the river-jungle; and in its *unstratified*, the eruptive mass, the molten overflow, and the frequent shower of dust and ashes. Interrogated as mere rock-masses, they expand overflow after overflow, and stratum upon stratum, like the leaves of a mighty volume, and tell of gigantic rivers and estuaries, of shallow seas, tides, and ocean-currents, of low-lying continents and volcanic archi-pelagoes, of shell-beds and coral-reefs, of vegetable growth and vegetable drift, of rains that fell, winds that blew, and suns that shone and gladdened the face of nature even as they do now. Of all this, and much more, these coals and sediments bear abundant testimony; and interesting as it must ever be to the educated mind to trace back the unity

---

* A curious proof of the morass or swamp-growth of many of our coal-seams, is to be found in the narrow winding "wash-outs" by which they are frequently intersected. These "wash-outs" of the miner are stream-like courses from which the coaly matter has disappeared, its place being taken by stony substances. They have been clearly runnels or water-courses that threaded their way through the swamps, and thereby prevented the accumulation of the vegetable matter, just as at the present day our peat-mosses are cut into channels by the streams that may drain their surfaces.

and continuity of the physical agencies that mould and modify the face of nature, that interest becomes immeasurably
enhanced when we associate the results of these old-world
operations with the necessities of the present, and trace in
them an obvious provision for the social and intellectual
advancement of man.

We come next to consider the Carboniferous system as a
life-period, and though there must necessarily be considerable differences between the fossils of its respective divisions—carboniferous shales, mountain limestone, and coal-
measures—yet in a sketch of this kind the aim is more
an outline of the whole than the consideration of specific
minutiæ, which can only be appreciated by the professed
palæontologist. Perhaps the most remarkable feature of the
period is its *Flora*—a flora remarkable not only for its vast
exuberance, but for the peculiar character of its plant-forms,
which bear, in most instances, but a faint resemblance to
those of the present day. This vegetation is for the most
part converted into coal, but here and there, scattered
throughout the shales and sandstones, we find leaves, fruits,
stems, trunks, and roots, which indicate its nature, and
from these the botanist must construct the aspects of the
carboniferous flora. Sea-weeds, marsh-plants like the equisetums, reeds, and rushes, a vast variety and exuberance of
gigantic ferns and clubmosses, pine-like trees with their
leaves and cones, and a still greater number perhaps which
cannot be assigned to any existing order, may be said to
constitute the bulk of the coal vegetation. Fragmentary,
and converted into coaly or stony matter, the botanist has
no easy task in reading these old-world forms, and all that
he can in many instances do, is to trace a resemblance and
give a name founded on some external peculiarity. It is
for this reason that we find in lists of carboniferous plants

such names as *calamites* (reed-like), *equisetites* (equisetum-like), *lycopodites* (clubmoss-like), *sphenopteris* (wedge-leaf fern), *neuropteris* (nerve-leaf fern), *lepidodendron* (scaly-bark tree), *bothrodendron* (pitted tree), and so forth, all pointing to some obvious feature which distinguishes them one from another, but throwing very little or any light on their true botanical affinities. But whatever their strict affinities, we know that most of them belonged to the lower orders of vegetation—the horsetails, ferns, and clubmosses, the grasses, sedges, and rushes, the cycads and pine-trees, or perhaps more properly to extinct forms that stood intermediate, as it were, between these various orders. Though lowly in organisation—and most of them were undoubtedly so—they seem to have occupied large areas of the earth for ages, and to have grown in rank luxuriance, till in numerous instances their accumulated masses form seams of coal from a few inches to many feet in thickness.

When we turn to the *Fauna* of the period, we find throughout the same variety and the same numerical abundance, though, of course, certain forms are more abundant in one portion of the system than in another. These forms, too, are chiefly aquatic—fresh-water, estuarine, and marine; there being few terrestrial species yet discovered, and these only at wide intervals and in few localities. Beginning with the lower forms, we have a number of minute foraminiferal organisms, and a vast exuberance of corals and encrinites, so vast that beds of the mountain limestone, hundreds of feet in thickness, are almost entirely made up of their remains. There are also trails, burrows, and tubes, that indicate the existence of marine annelids; abundance of crustaceans, some minute and bivalved like the cypris, a few species of trilobites, and others like the king-crab, and of large dimensions. The polyzoa or flustra-like organisms occur too in great variety, scattering their netted cells through the shales and limestones; and shell-fish of every known order — bivalve and univalve, deep-sea and shore dweller—occur throughout the entire system, though most abundantly, of course, in the marine beds of the mountain limestone. Fishes of many forms are likewise abundant, especially in the lower series of the system, their shining enamelled scales, predaceous teeth, and defensive fin-spines being scattered through the shales, ironstones, and limestones. Many were large and shark-like, their palatal teeth, jaws, scales, and fin-spines indicating lengths from twelve to eighteen and twenty feet, and bulky in proportion. Being chiefly cartilaginous, their bodies have in most instances utterly disappeared, and only their teeth or enamelled fin-spines remain to testify to their existence. We have seen hundreds of teeth and spines from a single layer of black-band ironstone, and yet no other vestige of the fishes to which they belonged, not even a patch of scales in juxta-position, to indicate their affinities. Higher than fishes, reptiles also make their decided appearance, most of them aquatic and fish-like in form, though a few ascend to true lacertilian or terrestrial species. Of the terrestrial fauna of the period we know little; but the insects, land-snails, and reptiles of arboreal habits, which have been found in certain coal-fields, were surely not the sole inhabitants of the carboniferous islands and continents, and we may safely look forward to the discovery of other and higher forms of which these were the necessary congeners. Even while we write, the announcement of several new genera of reptiles from the coal-field of Kilkenny gives additional encouragement to this expectation, and all that seems necessary to its fulfilment is merely more minute research and more careful examination on the part of palæontologists. Indeed, when we consider the difficulty of preserving terrestrial organisms, how much they are subjected to waste and decay, and how few are necessarily washed down into estuaries and seas of deposit, it is wonderful to learn that such fragile remains as those of insects, land-shells, and tree reptiles should have been saved from destruction. And surely, if larger and stronger forms had existed, the hope may be indulged that they too have been preserved, and will one day or other be detected.

Such is a hurried glance at the Life of the old Carboniferous period, and more especially as displayed in the areas of Europe and North America, where mining operations have been most extensively conducted. Whether these extensive coal-fields were all contemporaneous is a subject open to question. Indeed, the probability is that they were not strictly contemporaneous, but merely belonged to a great cycle of the earth's history characterised by these coal-forming conditions, and in the main by the sames facies of plants and animals. But however this may be decided by future and more exact inquiry, we perceive in the mean time a wonderful similarity all over the old coal-measures, and an exuberance of life that has never been excelled during any subsequent epoch. To account for this exuberance, especially in the vegetable world, various hypotheses have been advanced, such as a greater proportion of carbonic acid in the atmosphere, the greater amount of heat derived in those earlier times from the interior of the globe, a general lowering of the land-surfaces, and a higher temperature over the coal-yielding areas arising from some change (inclination of the earth's axis or otherwise) in the astronomical relations of our planet. In the present state of our knowledge and belief in the stability of the earth's planetary relationships such hypotheses are inadmissible, and we are driven to seek for the solution in the then distribution of sea and land, the climate thereby produced, the nature of the vegetation, of which we as yet know too little, and the long continuance over the same areas of the same external conditions. So far as we can judge of the character of the vegetation (alliance to equisetums, clubmosses, tree-ferns, swamp-pines, and the like), it by no means required a tropical temperature for their growth and accumulation, but rather a moist, equable, and genial climate, inundated river-plains and morasses, low-lying deltas and sea-swamps—and these could be brought about by the terraqueous arrangements of the earth's own surface, and without calling in the aid of anything either preternatural or abnormal. We say, "could be brought about by the terraqueous arrangements of the earth's own surface," for it is not difficult to conceive such a *position* of the land-masses as to receive more heat from the sun and more warmth from oceanic currents, and such a *lowness* also of the terrestrial surfaces as to offer few points of condensation to aqueous vapours, and thus preserve a greater permanent amount of atmospheric moisture. This moisture would act in a twofold manner in promoting luxuriance of vegetable life—*first*, by affording a full and direct supply for their growth ; and, *second*, in lessening the radiation of heat from the land-surface, and thereby greatly increasing the general temperature.* But whatever the geographical conditions, they must have continued for long ages over the same areas to permit the accumulation of such a thickness of coals and sediments and igneous eruptions as those which constitute the Carboniferous system. And these accumulations imply vast continents from which they were wasted, large rivers for their transport, extensive deltas and sea-swamps for the growth of coal-beds, frequent volcanic eruptions in or near the

---

* For evidence of this peculiar effect of atmospheric vapour, see Professor Tyndall's reasonings and experiments in his ' Heat considered as a Mode of Motion.'

areas of deposit, and over the whole a gradual subsidence
to allow the depositions of bed above bed in such regular
and continuous arrangement. And while all this went on
the march of life was ever onward and upward. Plants
unknown in the Silurian and Old Red Sandstone periods
made their appearance; newer genera and species of corals,
shell-fish, crustacea, and fishes thronged the waters. Reptiles, so doubtfully known in the Old Red, now appeared
in considerable variety; insects, frail and fragile as they
generally are, were by no means uncommon; and all that
is wanting to complete the scheme of life, as now known to
us, is the presence of birds and mammals. Whether this
absence of birds and mammals arises from their non-ex-
istence during the period, or from the imperfection of the
geological record, it is impossible to determine; but clearly
the flora and fauna are greatly in advance of those of the
Old Red Sandstone, and all this is in perfect harmony with
the geological doctrine of a progressive development of
vitality.

As an economic repository the old coal-measures present
a wide field for inquiry and description. The variety and
value of their products, the skill and capital expended in
obtaining them, and their obvious bearings on the industrial and social relations of a nation, are subjects, however,
that lie far beyond the scope of a single sketch, and all that
we can attempt is little more than a mere enumeration of
the principal substances. We allude, of course, more especially to the Coal Formation of the British Islands, from
which it may be safely asserted that we derive products of
greater value than from all the other formations put together. From its *sandstones* we obtain many of our most
durable and beautiful building-stones; we fabricate its *fire-
clays* into furnace-bricks, retorts, drainage-pipes, baths, and

other articles of utility and ornament: from its *shales* we extract alum, copperas, sulphur, and paraffin oils ; its *limestones* are employed in architecture, agriculture, iron-smelting, bleaching, tanning, and numerous other arts, at the same time that they furnish many of our most decorative marbles, and are often the repositories of lead, zinc, antimony, and silver: from its *ironstones* we extract much of that metal without which all our implements would have been comparatively rude and inefficient, and the machinery of our factories, our steamboats, our railroads, and telegraphs impossible ; while with its various *coals* we heat our dwellings, cook our food, light our streets and apartments, and raise that steam-power by which human industry is increased ten-thousand-fold, time and space abridged, and the different nationalities of the earth brought into more intimate union and brotherhood.

And yet, important as these substances are, they are far from having attained their limit either in the amounts annually raised or in the purposes to which they can be applied. One has only to cast his eye back on the state of our coal-fields some thirty or forty years ago, as compared with what they are now, to be at once convinced of the progress that has been made, as well as of the progress that is still attainable. At that time, fire-clay was raised only from a few open workings, and had little or no value; alum-shale was mined only at a single work; blackband ironstone was rejected; bituminous shales were utterly worthless; the ordinary coals had for the most part merely a local sale (for there were no railroads), and brought less than half their present prices ; cannel coal was seldom raised, and was scarcely saleable even at a fifth of its present cost (for gas-works had not then come into operation); and the whole amount of coal raised in Great Britain did not much exceed 30,000,000 tons, while now it has reached

the enormous amount of 100,000,000, and is still steadily
increasing !   At that time steam-power had not come into
general use at our collieries, and horses, men, boys, and
women toiled indiscriminately under ground and above
ground.   Shallow workings, open inclines, and stair pits
were then the order of the day, with little profit to the
employer, and a world of discomfort to the employed.
Now our coal-works, though still demanding improvement,
are models of systematic engineering in comparison, and
are year after year conducted on a more enlarged scale and
on more strictly scientific principles, at the same time that
they have been placed under a humane and generous system
of official surveillance.

Looking, we say, at the progress that has been made
during the last thirty years, at the increasing demand and
pressure upon our coal-supplies, and at the general improvement in mechanical appliances that is incessantly taking
place, we may rely on increased national interest in all that
relates to our coal-fields, to new inventions for lessening
the toil and danger incident upon mining, and to a still
further utilisation of the substances that belong to the
Carboniferous system, and which are now only partially
employed or altogether neglected.   But while we look
hopefully forward to this utilisation, we cannot lose sight
of the fact that our coal-fields—we mean the British coal-
fields—have a limited area, and that at the present rate of
consumption a time must come when every available seam
will be exhausted.   How far distant this time may be our
best practical authorities are by no means agreed, some
restricting it to 300 or 400, some to 600, and others again
to 1000 years.*   It is true that by more economical methods

---

* Those interested in the probable duration of our coal-supplies may
consult Hull's ' Coal-fields of Great Britain,' the most compendious, as
well as the most readable, work on the subject to which we can refer them.

of consumption the present increasing demand will be somewhat restrained; that with improved methods of working much of the mineral now left underground will be raised; and that with more skilful engineering deeper shafts may be sunk through the overlying secondary rocks. But even with all these appliances no man of intelligence can shut his eyes to the facts that we are rapidly working out our best and most accessible coal-seams, that deeper winnings must entail greater expense, and that in less than a century hence the price of British fuel will be immensely increased. How long the supply may be sufficient to sustain the supremacy of British industry it is impossible at present to determine, but assuredly two or three hundred years hence all the more accessible portions of our coal-fields will be thoroughly exhausted, and our successors will be driven either to foreign fields, to other sources of heat than coal, or to other centres of industry. No doubt the vast fields of America and Australia are scarcely broken in upon,* and science is every year discovering newer fields in other regions, and this will tend greatly to lessen the pressure on those of Great Britain; but still the results must ultimately be a change in our commercial relations and a shifting of the theatres of manufacturing industry. Under a wide and cosmical view, however, such changes are inevitable, and we need no more disquiet ourselves about the future condition of our country than about the future distributions of the seas and continents. The rise and decline of nationalities and the phases of their commercial power

* The available coal areas of Great Britain are usually estimated at little more than 5000 square miles; those of North America alone exceed 200,000! When we add to this the unknown areas of South America, of Australia, of Japan, China, and India, to say nothing of the partially explored fields of Russia and Austria, the reader will readily perceive how immense the stores of fossil fuel laid up for the future requirements of human industry.

are as clearly under a higher law of progression as are the physical and vital appointments of the globe itself, and to the philosophical conception all great mutations are merely successive stages towards the broader and higher attainment.

But be this as it may, to our country these Old Coal-Measures have been invaluable : they are the mainspring of her mechanical power and the stay of her commercial greatness, and everything that tends to economise their use or prevent their unnecessary consumption should be hailed as a national advantage.   By their aid mankind has gained new triumphs over the powers of nature, and modern civilisation been infinitely accelerated.   One cannot, indeed, reflect for a moment on the utilities of this single formation, without discerning how intimately the most recent period in geology is connected with the most remote, and how all are inseparably woven into one beautiful and harmonious world-plan.   Strange, that the mere physical operations of the earth's remotest ages should be so intimately associated with the industrial and intellectual operations of the present !   Inexplicable, if creation were not the unfolding of a great *moral* as well as of a great *physical* scheme—a scheme in which every element and operation in time past as in time present plays an essential part, and from which not a jot or tittle could be abstracted without marring the symmetry and perfection of the whole.

WHAT WE OWE TO OUR COAL-FIELDS.

BRITAIN'S SUPREMACY IN MECHANICAL AND MANUFACTURING IN-
DUSTRY DEPENDENT ON HER COAL-FIELDS—PHASES OF MODERN
AS COMPARED WITH THOSE OF ANCIENT CIVILISATIONS—DIF-
FERENCES ARISING CHIEFLY FROM THE USE OF COAL AND
IRON—SPECIAL PRODUCTS OF OUR COAL-FIELDS—COAL AND ITS
VARIETIES—IRON AND THE AGE OF IRON—LIMESTONES AND
MARBLES—SANDSTONES AND THEIR RELATIONS TO ARCHITEC-
TURE—FIRE-CLAY AND FIRE-CLAY FABRICS—SHALES, AND THE
EXTRACTION OF ALUM, COPPERAS, PARAFFIN, AND PARAFFIN OILS
—ORES OF LEAD, ZINC, AND SILVER—RELATIONS OF MECHANICAL
AND MANUFACTURING INDUSTRY TO COAL AND IRON—RELATIONS OF
INDUSTRY AND COMMERCE TO CIVILISATION AND REFINEMENT.

EVERY man of thought must be more or less impressed
with the conviction that much of Britain's supremacy in
mechanical and manufacturing industry has arisen from her
rich and readily-accessible coal-fields. A high degree of
civilisation, as the histories of ancient nationalities demon-
strate, may be attained without the possession of coal-
fields; but the peculiar phases of civilisation, in all that
relates to mechanical appliances, manufactures, locomotion,
and intercommunication, are the direct results of coal and
iron. The fine arts, literature, philosophy, social refinement, and political institutions have existed, and may yet
exist, where coal-fields are unknown; but that *machine-
power*, if we may so express it, which coal and iron put
into the hands of man to subdue the forces of nature, and
thereby promote the wider advancement of his race, intel-

M

lectually as well as materially, is a thing dependent alone
upon the existence of a Coal-formation.   There is no artificial source of heat (and heat is the spirit of all force) so
compact, so portable, so safe, and so readily available as
coal; no substance so adaptive, so strong, and so enduring
as iron.   There is no artificial power so titanic, and withal
so submissive and tractable, as a few pounds of ignited
coal acting through the medium of water; no harness
save one of iron sufficiently strong to yoke that giant power
in the services of human industry.   These two substances,
coal and iron, have been the main factors in all recent progress; and that which most broadly distinguishes the
Britain of the present from the Britain of the preceding
centuries is the extended and extending use of these substances through the instrumentality of the steam-engine.
Nor is it for these two minerals alone that we are indebted
to the Carboniferous system; for from the same formation
we obtain numerous other products, all useful in the arts,
and some of them indispensable auxiliaries to the employment of coal and iron.   These are the sandstones, limestones, marbles, fire-clays, oil-shales, alum-shales, and cop-
peras-shales, and not unfrequently in some localities the
ores of lead, zinc, and silver.

Taking COAL as the chief product, we find it in many
varieties and in all degrees of purity:—*Caking coal*, like
that of Newcastle, which is tender and bituminous, and
cakes together in burning; *cubic* or *rough coal*, which is
harder, burns open, and leaves much ashy residue; *splint*
or *slate coal*, like that of Fife, which is hard and laminated,
and burns open, with the emission of great heat; and
cannel or *parrot coal*, compact and jet-like in texture, rich
in bitumen, and now chiefly used in our gas-works.   One
or other of these varieties is found in every coal-field, and
according to its quality (for, like all mixed rocks, coal is

very variable in quality) is fitted for some special purpose—domestic fuel, steam-raising, gas-making, oil-making, or metallurgy. For these and kindred purposes about 100,000,000 tons are now annually raised from British coal-fields; and when we reflect on the necessity for fuel in our northern climate, on the advantages of a cheap and convenient light like gas, on the importance of our manufacturing machinery which is moved by steam-power, and on the impulse which steam has given to our intercommunication by land and sea, it is impossible to over-estimate the importance of this single mineral. In young and newly-settled countries the forests may supply its place; but as population increases, as the land must be cleared for grain crops, and as the forest-growth necessarily disappears, coal is yet the only available source of heat, and happy the nation that possesses it in accessible abundance! To Britain it is all-important; and when we consider the amount of capital, skill, and industry employed in procuring it, the myriad economical purposes to which it is applied, the facility with which it can be carried, and the safety and readiness with which it can be used, we may well wonder what would have been the condition of our country without it. This, then, we owe to our Coal-fields; and considering the enormous amount raised from so small an area, the rapidly increasing demand, and the undoubted limits of the supply, everything that tends to facilitate its full and careful extraction or economise its use should be hailed as a boon of no common importance.

We have alluded to the enormous amount of coal raised from so small an area, and the aggregate of the coal-fields of Britain, though large in comparison with what is possessed by other European countries, does not greatly exceed 5000 square miles. Even in this area much of the coal is of inferior quality, much is also unworkable, and a considerable amount

must always be left in the process of extraction.   Under
these circumstances, and with a consumption that has
trebled itself within the last thirty years, it is not to be
wondered that some uneasiness has been felt as to the duration of our coal supply, and a Royal Commission been
appointed (1866) to inquire into the probable amount that
still lies unworked, and which seems to be fairly accessible.
Whatever be the issue of the inquiry, two things are certain
—*first*, that the supply is limited, and that at the present
rate of consumption it will be wholly exhausted in a few
hundred years ; and, *second*, should the coal-measures be
found to extend under the newer formations, it would be impossible, with our present appliances, to extract the mineral
at these enormous depths, and even if engineering skill were
enabled to surmount the difficulty, the increased cost would
be tantamount, in all ordinary cases of consumption, to
closing the supply.   In the mean time, and without dwelling on this most unwelcome prospect, our duty is clearly
to encourage every plan for the fuller and more careful extraction of the mineral, and to do what we can, individually
and nationally, to economise the consumption.

Regarding IRON as the product of next importance, it
may be said to occur in two varieties in our coal-fields—
namely, as a rough clay-carbonate in bands and nodules,
and as a finer clay-carbonate in beds of varying thickness,
and mingled less or more with coaly or bituminous matter.
The former constitutes the " clay-band " of the miner, and
the latter the " black-band "—the former being the more
abundant, the latter the more valuable as containing enough
of coaly matter for its calcination without the addition of
extraneous fuel.   There are other and more abundant sources
of iron than the coal-formation (the hæmatites and siderites) ;
but, considering that these ores cannot be reduced to the
metallic state without the addition of fuel, the conjunction

of ironstone and coal in the same formation—in the same mine, we may say—renders these clay-bands and black-bands of peculiar importance. About 5,000,000 tons of pig or cast iron are annually manufactured in Britain, and of this the greater proportion is obtained from our coal-fields. Looking at the peculiar phases of modern civilisation, it is impossible to over-estimate the importance of an abundant supply of iron. Indeed, without this abundance, our varied machinery, our steam-engines, our railroads, gas and water pipes, suspension bridges, iron ships, and other similar inventions, would have been impossible, or if possible, would have been on a limited and insignificant scale in comparison. This work and application of iron is one of the leading features of our times—its malleability, strength, durability, and cheapness fitting it for almost every purpose, from the homeliest utensil of domestic use to the most elegant and complicated machinery, and from the tiniest implement of art to the most gigantic instrument of war. Ours is in truth the *Age of Iron*, there being no industrial operation in which that metal does not play an important part, no enduring subjugation of the forces of nature unless through its instrumentality and power.

Besides coal and iron, LIMESTONE also takes rank as one of the abundant and valuable products of our coal-fields. Associated with coal and ironstone, it becomes of especial value as a flux for the reduction of the latter in the blast-furnace, and all the more valuable that it is obtained from the same formation, and generally in convenient proximity. It is also extensively used for mortar and cements, in agriculture, in tanning, bleaching, and other industrial processes, and not unfrequently some varieties are hard enough and beautiful enough in texture to be raised and polished as marbles. Taken together, there are no mineral substances so valuable as coal, ironstone, and limestone; and it seems something more than a mere coincidence that in the coal-formation they should have been associated in such close proximity and accessible abundance. Had there been no other rocks in the carboniferous series save these three, the gifts of our coal-fields would have been incalculable; how much more their value when other products, all useful and abundant, can be raised from the same system, from the same shaft, and in course of the same mining operations!

SANDSTONES of great beauty and durability, like those of Edinburgh, St Andrews, Stirling, Glasgow, Newcastle, Leeds, and other localities, are obtained in inexhaustible supplies from the Coal-formation; and when the importance of substantial and elegant edifices is duly considered, these building-stones must be ranked among the more valuable of its products. Taking Edinburgh, Glasgow, and New-castle as examples, the fitness, beauty, and durability of the coal-measure sandstones for architectural purposes must stand unquestioned; and, considering the readiness with which they can now be carried by railway to all parts of the island, the rapid extension of their employment may be safely predicted. Nor is it merely for building purposes that these sandstones are used, but many of their varieties, as their names imply (millstone grit, grindstone grit, &c.), are extensively raised for millstones, grindstones, whetstones, and other kindred purposes. In a rigorous and irregular climate like that of Britain, the possession of a durable building-stone is of prime importance, alike on the ground of comfort and economy, and luckily the coal-formation in one or other of its series affords a cheap and inexhaustible supply. It was the boast of Nero that he found Rome built of bricks, but had left her of marble; had her seven hills contained such sandstones as those of Craigleith and Binny, the boast might have taken a different direction. Our own Houses of Parliament, whose decay so soon after

their erection has excited so much comment, are built of
magnesian limestone; had the sandstones of Fife or Mid-
Lothian been adopted, as was at one time proposed, the
chisel-marks of the builder would have remained unobliterated for centuries.

Next to the sandstones or freestones of the Coal-measures
we may place the FIRE-CLAYS, or those argillaceous beds which,
from the peculiarity of their composition, may be baked into
fabrics of any shape, and of unsurpassed resistance to the
effects of either fire or water. These clays, at one time
little sought after, have now become of vast importance,
and are largely used as substitutes for stone and iron.
Plastic, and capable of being fashioned into any form, they
are extensively used for fire-grates, furnace-linings, retorts,
sewage - pipes, floorings, architectural mouldings, garden
ornaments, and a thousand other purposes — affording
objects of beauty as well as of utility, and, when properly
manufactured, of unconquerable durability. Few manufactures have come of late years so largely into use as
fire-clay fabrics ; and, considering their recent introduction,
we may confidently look forward, not only to their rapid
extension, but to a skill in manipulation which will rival
the finest productions of the potter and sculptor. But,
altogether apart from works of art, the use of fire-clay
fabrics in furnaces, and above all in the cheap and efficient drainage of our towns, has been a boon of no common
value, as tending at once to the health, the comfort, and
amenity of our densely congregated populations. Of course,
as the manufacture of fire-clays is intimately dependent on
a cheap supply of fuel, the relation of coal and fire-clay in
the same field, and often in the same mine, is a thing that
must strike the attention of the least reflecting.

Even the most worthless-looking beds in the formation—the SHALES, or consolidated muds—are now rapidly

rising in economic importance. Those yielding alum and copperas (the alum and pyritous shales of the miner) have long been worked, though on a small scale, but now the bituminous shales, or those yielding paraffin and paraffin oil, are eagerly sought after and worked on a gigantic and rapidly extending scale. Fields in the lower Coal-measures of Scotland, which half-a-dozen years ago did not bring a farthing to their proprietors, are now yielding thousands; and the distillation of these shales may already be regarded as an established branch of our national industry. Nor is it merely as sources of a new and brilliant light that these shales acquire their importance. Recent experiments have proved the adaptability of their crude oils as a fuel for steam-raising, and the hope is now held out that even the inferior varieties may be turned to good account in lessening the pressure upon our more precious coal-seams. Indeed, when we consider the worthless aspect of these shales and the beauty and utility of the substances—solid paraffin, paraffin oil, naphtha, rosine and magenta dyes— derived from them, few instances of human ingenuity to utilise the products of nature could be adduced, at once so marvellous and so thoroughly successful. What so unlike as a block of black bituminous mudstone and a paraffin candle, white and translucent as the finest wax? What so seemingly impossible as the extraction of a brilliant rose-purple dye from a mass of pitch-coloured coal-tar?

Besides these rocks—the coals, ironstones, limestones, sandstones, fire-clays, and shales—of which the Carboniferous system is entirely composed, there occur in some localities independent veins of lead-ores and zinc-ores traversing the beds of the mountain limestone. These veins (like those of Derbyshire, Yorkshire, and Northumberland) are often of great commercial value, not only for the lead and zinc they directly supply, but for the percentage of silver which many of them yield—the lead-ores being generally less or more argentiferous.

Such are the products we derive from the Coal-formation, and such the benefits they have conferred and are still conferring on our country. Statisticians may set down their amounts in weights and measures and their values in money, but no method of computation can fully convey an idea of the advantages, physical, intellectual, and moral, we enjoy in possessing such productive and accessible coal-fields. Had the coals been in one field, the ironstones in a second, and the limestones and fire-clays in a third, they would have still been prized and sought after; but when the whole are found in the same field, and often in the same mine, their value is a hundredfold enhanced, and their importance becomes more distinctly appreciable. Nor can it be overlooked that, situated as most of these coal-fields are, in the lower districts of our island, they are peculiarly accessible by water and readily reticulated by railways—the Bristol Channel, the Mersey, the Tees, Wear, Tyne, Forth, and Clyde all opening their ports for convenient transport to other and less favoured localities. Nor is it to Britain alone that the advantages of her coal-fields extend—the whole world, through her machinery and manufactures, participates in the boon, and the impetus thereby given to human progress will descend to future ages. Her steam-engines, manufacturing machinery, railroads, water and gas distribution, telegraphs, and steamships are the giant offspring of her coal-fields; and wherever the sound of these is heard or their influences felt new activities and industries are awakened, and on industry and commerce are ever founded the surest hopes of civilisation and refinement.   Strange, as we have before remarked, that the mere physical operations of the earth's remotest ages should be so intimately associated with the industrial and intellectual operations of the present! Inexplicable, if creation were not the unfolding of a great *moral* as well as of a great *physical* scheme—a scheme in which every element and operation in time past as in time present plays an essential part, and from which not a jot or tittle could be abstracted without marring the symmetry and perfection of the whole.

THE SECONDARY AGES.

WHEN the earlier geologists arranged the rocks of the earth's crust into Primary or first-formed, Secondary or second-formed, and Tertiary or third-formed, they made a most important improvement in geological classification; and though a somewhat different meaning is now attached to these terms, they are still retained in the nomenclature of the science as general and convenient designations. Accepting PRIMARY merely in the sense of *early* or *ancient*, it embraces all the stratified systems prior to the New Red Sandstone; while SECONDARY, on the other hand, refers to the New Red Sandstone, the Oolite and Chalk. Even this New Red Sandstone, when critically examined, is found to differ widely in its fossils—the lower portion containing palæozoic forms closely related to those of the Coal-measures, while the upper portion is characterised by mesozoic forms, or such as have a closer relationship to those of the Oolite and Chalk. In this way the "New Red Sandstone" of our

forefathers has been separated into two distinct systems—
the "Permian" or Lower New Red, so called by Sir Roderick
Murchison because typically displayed in the province of
Perm in Russia ; and the "Triassic" or Upper New Red, so
named by the German geologists because consisting of three
well-marked series of strata, the Bunter or variegated sandstone, the Muschelkalk or shell-limestone, and the Keuper
or copper-marls.   Adopting this view—and it is more by
nature of the fossils than by the character of the sediments
that the relative antiquity of strata can be satisfactorily
determined—the secondary ages, which form the subject of
the present sketch, comprehend the Triassic, Oolitic, and
Chalk systems, whose chronological place will be best understood, perhaps, by a glance at the annexed tabulation :—

CAINOZOIC or TERTIARY.	Quaternary or Recent Accumulations. Tertiary.
MESOZOIC or SECONDARY.	Cretaceous or Chalk. Oolitic or Jurassic. Triassic or Upper New Red Sandstone.
PALÆOZOIC or PRIMARY.	Permian or Lower New Red Sandstone. Carboniferous or Coal. Old Red Sandstone and Devonian. Silurian. Cambrian. Laurentian.

These secondary systems—the Trias, Oolite, and Chalk—
hold a middle place, as it were, in geological history, less
obscure than the primary, but still not so obvious either in
their vital or geographical arrangements as the tertiary and
recent.   The mode in which their strata have been aggregated is for the most part apparent, and their fossils, though
differing widely in genera and species from existing forms,
have still more of a new-world aspect about them, and the
palæontologist feels he has less difficulty in establishing
their botanical and zoological relations.   Unless in highly

igneous centres, few of the strata have suffered much metamorphism; the areas and boundaries of the seas of deposit
are much more apparent; and birds and mammals, unknown
in previous systems, are now met with in some abundance.
Of course, as the secondary ages embrace several systems
and formations, there will necessarily be considerable difference, both in rocks and fossils, between its respective series;
but notwithstanding these differences there is still similarity
sufficient to enable us to treat them as a great group, or
rather as occupying a continuous and connected portion of
the earth's geological history. And after all, it is only by
groupings and generalisations of this kind that the non-
scientific reader can be expected to catch a glimpse of world-
history, the details and reasonings of which must be
worked out by the slow and patient research of the profes-
sional inquirer.

The earliest of these Secondary systems, we have said,
is the *Trias*, a series of reddish-coloured sandstones, shelly
limestones, and saliferous or salt-yielding marls; the second,
the *Oolite*, a series of calcareous freestones, clays, and
shales, with occasional coals and ironstones; and the third,
the *Chalk*, so notably composed in the south of England of
that white, soft, earthy limestone we call "chalk," and its
underlying clays and greensands. With the exception of
a subordinate series of strata that lie between the Oolite and
Chalk, and known as the " Wealden " from their occurrence
in the wolds or woodlands of Kent and Sussex, all the
strata of these secondary formations are eminently marine,
and are charged throughout with marine organisms. The
Wealden sands, sandstones, and clays point more to estuarine conditions, and where they occur the coals and lignites
of the Oolite and Chalk indicate the existence of land-surfaces; but all the other strata—the shelly and coralline lime-

stones and sandstones, the saliferous marls, the shales and clays * replete with the exuviæ of shell-fish, crustacea, and fishes—are strictly sea-formed, some *littoral* or along-shore, and others *pelagic* or in deeper waters. This marine origin is pre-eminently observable in the Trias and Chalk, and it is only in these oolitic districts where coal and ironstone are found that we have evidence of estuarine and terrestrial conditions of formation. Of course we are here alluding more especially to the secondary systems as displayed in

---

* A better idea of the composition of these systems may be obtained, perhaps, by a perusal of the annexed tabulation :—

CHALK or CRETACEOUS.	Chalky marls and thin-bedded limestones. Chalk rock, with and without imbedded flints. Greensands and cherty limestones. Blue tenacious clays, locally known as "gault" or "golt." Green and ferruginous sands, sandstones, and cherty limestones.
OOLITIC or JURASSIC.	Wealden clays, sands, and flaggy sandstones. Upper Oolite—consisting of fine-grained oolitic limestone with interbedded clays and bituminous shales (*coals*). Middle Oolite—consisting of coarse-grained shelly and coralline limestone, with blue clays and shelly calcareous grits. Lower Oolite—consisting of coarse shelly limestone and grits; thick-bedded blue clays (*coals*); thick-bedded sandy oolitic limestone; flaggy grits and marls; and calcareous freestone. Upper Lias—consisting of dark bituminous shales and indurated marls or marlstone. Lower Lias—dark laminated limestones and clays, bands of ironstone, layers of *jet* and *coal*, calcareous sandstones.
TRIASSIC or UPPER NEW RED.	Keuper—consisting of variegated marls, sandy layers, limestone layers, gypsum, and rock-salt. Muschelkalk—grey shelly limestone, with marl partings and beds of magnesian limestone. Bunter—soft variegated sandstones; coarse-grained grits and conglomerates.

Western Europe, for it must be observed that in India, Eastern Europe, in Farther Asia, and Southern Africa, as well as in various tracts of North America, the conditions of deposit seem to have been widely different, as there both oolitic and cretaceous coal-fields are by no means uncommon. And these coals, as noticed in a former sketch (Coals and Coal - Formations), must have been formed, like all others, partly from vegetation that grew and accumulated *in situ*, and partly from drifts and rafts borne down by rivers during seasons of flood and inundation.

There is thus nothing very puzzling in the lithology or mere rock-formations of the secondary ages. In the sandstones of the Trias studded with footprints and rain-prints, and reticulated with sun-cracks, we see the sandy deposits of shallow shores; in the shelly limestones, the accumulations of somewhat deeper waters; and in the clayey marls, with their masses of rock-salt and gypsum, the muds of lagoons and sea-creeks alternately submerged and cut off from communication with the outer waters. In the limestones and calcareous clays of the Lias and Oolite, we perceive the coral-growths and organic debris of exterior seas; in their freestones and shell-beds, the drifts of the nearer shore; while in their coals and jets we discover the growths of sea-swamps and deltas, and the drifts of streams and rivers. Again, in the greensands of the Chalk we trace the operations of the more exposed shores; while in the Chalk itself we perceive the slowly-accumulated calcareous ooze of the quieter waters. For just as in the Atlantic and other sea-beds, the fine calcareous mud resulting from organic debris (the shields of foraminifera, the waste of corals and shells and other exuviæ) is now collecting in great thickness and purity over extensive areas, so in the older cretaceous ocean similar agencies were at work in accumulating those masses of chalk which now constitute the cliffs and downs of southern England.* There is nothing, we repeat, very difficult of explanation in the mere sediments of the secondary ages, while, as during all other periods, vulcanic eruptions were here and there breaking up their continuity, and occasionally alternating with their strata. In England the secondary rocks have suffered little or no disturbance from igneous agency, and hence their broad and unbroken succession in that country. All to the south-east of a line roughly drawn from the Severn to the Tees is occupied by these formations, and the reader has only to cast his eye over the geological map, to perceive how regularly and continuously they follow each other. In Scotland and Ireland, however (Skye, Giant's Causeway, &c.), as well as in the Jura and other Continental districts, the secondary rocks are much disturbed and altered by igneous eruptions—a proof that during this, as during all other ages, vulcanicity displayed itself only along certain lines and within certain centres with notable intensity.

As a life-period, the secondary systems, though characterised each by its own peculiar forms, have yet so many features in common, that they may conveniently be regarded as representing one great and unbroken cycle of world-history. As marine deposits they abound in all the lower forms of life—foraminifera, sponges, corals, encrinites, star-fishes, sea-urchins, polyzoa, shell-fish, crustacea, and worms; contain at the same time many orders of insects; are replete with numerous families of fishes and reptiles; and now, for the first time in geological history, give unmistakable evidence of birds and mammalia. Without

* For more detailed explanation of this deep-sea calcareous ooze or mud, see page 44, in the sketch entitled "Waste and Reconstruction."

dwelling on minutiæ, which would be out of place in a popular sketch like the present, we may yet advert to some of the more remarkable features in the life of the secondary ages, such as the great preponderance of nautilus and cuttle-fish like forms among the mollusca; the marvellous variety of reptilian life, which has led to the designation "the age of reptiles;" the first unmistakable appearance of bird-life; and the occurrence of mammals of the lower or pouch-bearing section. There are other noticeable features too, such as the disappearance of the plants peculiar to the coal-measures, and their replacement by other tree-ferns, by cycads, zamias, pine-trees, and palms; the disappearance of the graptolites, trilobites, eurypterites, and bone-encased fishes that characterised the older systems; the prevalence of homocercal or equal-lobed-tailed fishes unknown in palæozoic times, and similar peculiarities; but these we must subordinate to the more remarkable features above alluded to.

The highest order of molluscan life is the Cephalopod, or those which move about by the arm-like feelers that encircle the head. To this order belong the nautilus and cuttle-fish—the former possessing an external chambered shell, and the latter having no external shell, but supported by an internal bone or osselet. The nautilus is now the only representative of the shelled division, and though a number of genera belong to the shell-less orders, they do not occur in anything like overwhelming numbers. In the secondary seas, however, these cephalopodous mollusca were predominating forms, their shells and internal bones occurring in myriads, and this more especially during the oolitic and cretaceous periods. Everywhere throughout the limestones, shales, and clays, their remains are scattered in profusion—the chambered shells, from the size of the smallest coin to the circumference of a carriage-wheel, and the internal bones from the thickness of a quill to the size of a man's arm. There are few things so noticeable in these secondary ages as this exuberance of chambered shells—*ammonites, baculites, hamites, scaphites,* and *turrilites;* or of these internal bones—*belemnites, acanthoteuthis, belemnoteuthis, conoteuthis,* and *leptoteuthis.** Many of the strata are surcharged with their remains; and as the cephalopods are active and predaceous in their habits, the lower life in which they fed must have been still more abundant, while they in turn became the chosen food of the larger fishes and reptiles. Occasionally we hear, as a matter of marvel, of the capture in tropical waters of cuttle-fishes eight or ten feet in length; how transcendently more wonderful the thronging of these secondary seas with myriads of the same order still more gigantic in size and more diversified in feature! Strange, too, that the ocean which then swarmed with nautiloid forms should now contain only a single genus! Inexplicable, were there no plan of vital progression to which these extinctions and creations could be systematically conformed.

Even still more remarkable was the exuberance of reptilian life that thronged the seas, the estuaries, the rivers, and river-plains of the secondary ages. Nothing known before or since of reptile life is at all comparable either in point of variety, size, or numbers. Our museums are replete with their remains, and in conditions so perfect, that almost every feature can be restored in natural proportion and life-like reality. Whale-like and confined to the waters, crocodilian-like and amphibious, mammalian-like and treading the land, or bird-like and winging the air—

---

* *Ammonites* (coiled up like the horn of Jupiter Ammon); *baculites* (straight or staff-shaped); *hamites* (hook-shaped); *scaphites* (boat-shaped); *turrilites* (tapering and turret-shaped); *belemnites* (dart-like); *acanthoteuthis* (thorn-like organ); *belemno-* (bolt-like); *cono-* (conical); *lepto-* (slender); and so on.

these saurians were undoubtedly the most notable features of the period. It was indeed the "age of reptiles," when every habitat—aquatic, terrestrial, and aërial—was occupied; and every function—carnivorous, herbivorous, and insectivorous—was performed by these creatures. Frog-like, but bulky as an ox, the *labyrinthodon* squatted on the muddy river-flats; whale-like, the *ichthyosaur* paddled through the waters; iguana-like, but huge as an elephant, the *iguanodon* browsed on the succulent herbage; bloodthirsty and mightier still, the *deinosaur* crouched for his prey in the forest; while bat-like, under the cliffs and adown the river-banks, fluttered the insectivorous *pterodactyle.** Every place seems to have been usurped and every function performed by these abounding reptilian orders; and this for long ages, while bird-life was gradually coming into force, and mammalian life dawning into existence.

This bird-life makes its first appearance in the Trias, the surfaces of whose flaggy sandstones are thickly imprinted with their footsteps. These *ornithichnites*, or bird-footprints, occur in all sizes, from the slender impress of feet like those of the snipe and sandpiper to the heavy implant of others many times bulkier than those of the largest ostrich. Waders and runners, they seem to have frequented the sandy shores of creeks and estuaries which were also the favourite haunts of reptiles, and thus many of these Triassic sandstones are crossed and re-crossed, tracked and re-tracked by footprints whose nature (reptilian or ornithic) it is yet impossible to determine. How wonderful the variety and complexity of the vital record! how strange that the impression of a passing foot should remain as evidence of scenes of life and activity when every other relic

* *Labyrinthodon,* from the labyrinthine structure of its teeth; *ichthyosaur,* fish-like saurian; *iguanodon,* having teeth like the existing iguana; *deinosaur,* terrible saurian; *pterodactyle,* finger-winged or flying reptile.

has utterly disappeared ! But numerous as bird-footprints are—and they occur in abundance at Storeton in Cheshire, Corncockle in Dumfriesshire, Cummingstone in Morayshire, Hildburghausen in Germany, and on the Connecticut in America—no traces of bird-bones have been detected save in a single instance, and that not altogether free from doubt, in the sandstones of the Connecticut. But what is doubtful in the Trias becomes obvious in the Oolite and Chalk, and the comparatively recent discovery of the skeleton of the *archæopteryx* * (ancient bird) in the lithographic limestone of Solenhofen confirms the fact that Bird-Life, whether existing or not in the primary periods, became at all events an established feature in the secondary ages. Yet so it ever appears with the great scheme of vitality ; advancing slowly but incessantly, and displaying at every stage more complicated forms and higher functional activities !

Conformable to this progress, mammals also make their appearance during the secondary ages—scantily in the Trias, but more abundantly and unmistakably in the Oolite and Chalk. As might be expected, and in accordance with a progressional scheme, those mammals belong to the lower or marsupial orders—that is, to those which, like the opossum and kangaroo, are furnished with a *marsupium*, or external pouch in which they carry about their immature young. These marsupials or pouch-bearers stand lower in the scale than the true mammals that bring forth their young in a perfect state. They are sometimes termed *ovo-*

---

* This ancient bird, according to Professor Owen, was about the size of a rook, and differs from all known birds in having two free claws belonging to the wing, and also in having the vertebræ of the tail (about twenty in number) free and prolonged as in mammals—each vertebræ supporting a pair of quill-feathers which give to the tail a long and vane-like appearance. This unique specimen (now in the British Museum) exhibits in its tail a retention of structure which is "embryonal and transitory in the modern representatives of the class Aves, and consequently a closer adhesion to the general vertebrate type."

viviparous, and, so far as this function of gestation is concerned, hold an intermediate place, as it were, between birds and the ordinary mammalia. Numerous teeth and jaws of a small size (about that of a rat or rabbit) have been discovered, some indicating herbivorous, some carnivorous, and others insectivorous habits, but all apparently belonging to the same pouch-bearing orders. It is true that in the chalk marls some bones of a doubtfully higher character have been detected, but beyond these fragments nothing higher in the scale of Life than birds and marsupials is known to belong to the secondary ages.

Here then, during these secondary ages, which embrace a long period of the world's history, we have seas of moderate depth and broad shallow estuaries in which all the ordinary sediments were deposited—shelly and coralline limestones in the outer waters, muds and clays in the stiller recesses, and sands on the open shores. Here and there masses of rock-salt and gypsum accumulate in detached lagoons and sea-reaches; while during the oscillations of the land the swamp and forest growths of a genial climate are repeatedly submerged and converted into coal. We say *genial climate,* for the cycads, zamias, palms, tree-ferns, and broad-leaved pines (which are the prevailing forms), point to warm and equable conditions; while the frequent oscillations of the land are amply shown in the numerous seams of coal and "dirt-beds" or ancient soils on which the forest-growths had flourished for ages.* And while these genial conditions prevail over sea and land, both are equally exuberant with

* One of the best known examples of an oolitic soil is the celebrated "dirt-bed" of Portland—an earthy, carbonaceous mass, replete with the roots and prostrate trunks of cycads, zamias, and other vegetation characteristic of and peculiar to those secondary formations. In fact, a genuine forest-mould, with its rotted leaves, fallen trunks, an mbedded root-stools.

life—the former teeming with foraminifera, sponges, corals, encrinites, starfishes, shell-fish of every order, fishes and reptiles; and the latter with gigantic ferns, reed-like grasses, cycads, zamias, palms, and pine-trees, which became the chosen food or shelter of numerous insects, of reptiles terrestrial, arboreal, and aërial, of birds, and of marsupial mammals. What a busy panorama of life, growth, and decay is presented by these secondary ages! And not alone mere growth and decay, but development and progress; for during the long periods that elapsed between the commencement of the Trias and the close of the Chalk, thousands of forms became extinct—newer and higher ones taking their places, and at every stage approaching nearer and nearer to those of the Tertiary and Current epochs. So perceptible indeed is this approach that it has been proposed to arrange the geological formations into two great divisions only—the *palæozoic*, embracing all to the close of the Permian; and the *neozoic*, all from the commencement of the Trias up to the present day. But whatever may be the value of such arrangements, the fact of progression is everywhere obvious —so obvious, that even the non-scientific observer could have little difficulty in distinguishing between the fossil forms of the primary and secondary ages.

Let those who refuse their assent to a plan of vital development only study for a day or two the magnificent collection of secondary fossils in our National Museum, or even peruse their figures as given in ordinary geological works; and if at all capable of comparing, they cannot fail to perceive the vast advance that has been made upon the primary or palæozoic forms. Not only is there the introduction of birds and mammals formerly unknown, but the stamp impressed upon all the lower orders—corals, molluscs, crustacea, fishes, and reptiles—is that of greater complexity and specialisation. Let them next compare them with tertiary and existing forms, and note in these not only the further introduction of the true timber trees and the higher birds and mammals, but also the still greater degree of specialisation both in form and function which characterises the whole ; and unless blinded by prejudice or incapable of discernment, they cannot refuse assent that there has been a progress, and that this progress has taken place in a way so connected and definite as to lead to the unavoidable conviction of an all-pervading law of development.

As Economic Repositories the secondary formations are year after year assuming greater importance. Not many years ago, building-stone, limestone, lithographic slabs, fuller's earth, rock-salt, gypsum, and chalk, were the principal products extracted; but now ironstone like that of Cleveland in Yorkshire, coal from many fields (Austria, India, Indian islands, Brazil, Virginia, Vancouver, and other regions), coprolites or phosphatic nodules for manure, hydraulic cements, and other substances are largely obtained, and as foreign surveys are extended, will in all likelihood be met with in still greater abundance. The geological mapping of the world, by competent authorities, is merely in its infancy; * but as this proceeds the secondary systems are gradually revealing a larger amount of economic treasures,

---

* The survey of our own islands, though commenced many years ago, is not half completed ; and those of our colonial possessions—Canada, the West Indies, India, Australia, and New Zealand—are merely in their first stages. The same may be said of other European countries, most of which have made some progress with their surveys, but still feel the want of more minute and accurate information. The American States survey, begun with great ardour, has been finished only in a few instances ; while with the geology of South America, Arctic America, Asia, and Africa, we have the slightest and most imperfect acquaintance. Till all these areas have been more thoroughly explored, it were presumptuous to dogmatise, and idle to speculate, either as to the scientific aspects or the economic importance of the several stratified systems.

and exciting an interest more in keeping with that which has so long been attached to their palæontological and purely scientific aspects. In corroboration of this we need only refer to such discoveries as the Cleveland ironstone, which within the last dozen years has wrought such a revolution on the aspects and industry of northern Yorkshire; to the deposits of rock-salt near Middlesborough on the Tees and in Antrim; and, above all, to the fact that the main coal-fields of India and the East are of secondary origin. How changed the aspect of opinion within the last thirty years, when our predecessors, generalising from their limited knowledge, regarded coal and iron as belonging alone to formations of earlier date, and sought for traces of their existence only in connection with these primary systems! But so it ever is; the more limited our acquaintance with nature and nature's operation, the more restricted our notions of her bounties, and the less prepared are we to avail ourselves of their benefits and amenities. It is only when mankind have taken the wider survey that they can arrive at sounder conceptions, and forego the conclusions they had drawn from their local and circumscribed experience.

TERTIARY TIMES.

WHEN the earlier geologists arranged the stratified rocks of
the earth-crust into Primary, Secondary, and Tertiary, they
closed the Secondary with the Chalk, and regarded as *Tertiary* all the sediments that occur above that formation. In
their estimation Tertiary strata were comparatively limited,
and they had no conception of the extent, thickness, and
variety of sediments, or of the long periods which this
thickness and variety of deposits must necessarily imply.
To them primary was equivalent to universal; secondary
was still very extensive, though not universal; but the tertiaries were mere local patches occupying shallow depressions in the older formations. As research went forward,
however, modern geologists began to discover that tertiary
strata differed widely among themselves both in composition and fossil contents, and that as a whole they were
entitled to be ranked in schemes of classification as a *system*
of sedimentary deposits. It was also found that while all

these sediments, or rather formations of sediments, were
highly fossiliferous, they were covered over, at least in the
greater portion of the northern hemisphere, by thick masses
of bouldery clay and gravel, devoid, or all but devoid, of
fossils; and this bouldery clay was considered as necessarily
limiting or closing the system. In this way the Tertiary
of the earlier geologists became to be broken up into *Tertiary* and *Quaternary*—the former embracing all the strata
that lie between the Chalk and Boulder-Clay, and the
latter the boulder-clay and all those superficial accumulations that have since been formed, or are still in process of
formation, by the ordinary agencies of nature. Tabulating
this arrangement, we have the

> QUATERNARY, or POST - TERTIARY, embracing all the
> superficial clays and gravels, the peat-mosses, swamp-
> growths, coral-reefs, lake-silts, 'estuary-silts, and sea-
> silts, with the boulder-clay beneath ; and the
> TERTIARY, all the regularly stratified clays and gravels,
> marly limestones, gypsums, and lignites or brown-
> coals, that lie between the Boulder-Clay and the Chalk
> formations.

In other words, the Quaternary or Post-Tertiary embraces
the rocks and records of the current epoch ; the Tertiary,
the rocks and records of a period that long preceded. And
this not a brief period, but one of long continuance—so
long, that during its currency there were many oscillations
of sea and land, many extinctions of genera and species,
and many introductions of other and newer races.

As might be expected in a system embracing a long
period of time, the rocks and fossils of the earlier tertiaries
differ considerably from those of the later ; and hence their
familiar arrangement into lower, middle, and upper ; or

into *eocene, miocene,* and *pliocene,* if we take, as Sir Charles Lyell has done, the relative proportions of recent and fossil shells which occur in the successive stages.* But whatever the nomenclature we adopt, it is clear that the clays, sands, gravels, marls, limestones, gypsums, and lignites which constitute the tertiary system are the varied sediments of lakes, estuaries, and shallow seas; and that in general the extent and boundaries of these areas of deposit are more apparent than those of the earlier systems. Indeed, many of the tertiary deposits occupy limited spaces in the existing continents, and thus we are led to infer that something like the present distribution of sea and land had then begun to prevail. At all events, much of the existing dry land was then above water, and supplied the material for the sediments, as well as the habitats for the terrestrial flora and fauna of the period. Of course, it is not contended that the continents had then assumed their present outlines, nor that many lands then existing are not now submerged ; but it is indicated that there is considerable relationship between Tertiary and Recent times, and that the fossils of the Old World tertiaries are more akin to the living plants and animals of the Old World than to those of the New ; while, on the other hand, the tertiary fossils of the New World bear a striking resemblance to the plants and animals that still flourish there. In other words, we are approaching in tertiary times more nearly to the existing ordainings of nature, and may therefore expect a closer

---

* Taking a hundred shells from the lower, a hundred from the middle, and a hundred from the upper tertiaries of the London and Paris basins, Sir Charles found that only a small percentage (3 to 5) of existing species occurred in the lower, hence *eocene* (Gr. *eos*, dawn; *kainos*, recent), or dawn of existing things ; that the number of existing species was somewhat less than the extinct (25 to 40) in the middle, hence *miocene* (*meion*, less), that is, less recent than the existing ; and that in the upper the existing species exceeded the extinct (70 to 90), hence *pliocene* (*pleion*, more), that is, more recent than the middle or lower divisions.

resemblance in all the phenomena, physical and vital, than it was possible to trace between the present and any of the remoter epochs.*

As Rocks there is nothing difficult to comprehend either in the nature or composition of the tertiary sediments. In one *basin* or area of deposit we have alternations of sands, gravels, clays, and lignites; in another, sands, gravels, clays, limestones, gypsums, and lignites; and in a third, clays, lignites, marls, and interstratified overflows of lava or showers of volcanic ashes. All these alternations are well displayed in the tertiary basins of London, Hampshire, Paris, Auvergne, the Lower Rhine, and Vienna; and the observer has no greater difficulty in comprehending the nature of these successions than he has in interpreting the sediments of any lake or estuary of the present day. In some basins the limestones may be hard and compact, or even siliceous, like the burr-stone of Paris; in others the sands may be consolidated into sandstones; in some, the lignites may be peaty and woody, while in others they are scarcely distinguishable from ordinary coal; but taking them all in all, the tertiary strata present few difficulties either as regards composition or the agencies concerned in their formation. In some of the areas of deposit, as evidenced by their fossils, the strata are strictly *marine*, in others they are fresh-water or *lacustrine*, and in some, again, there is an admixture of sea and river silts, which are consequently regarded as *estuarine* or *fluvio-marine*. Just as at the present day such seas as the Adriatic, Euxine, and Caspian are receiving their

* The reader who will take the trouble to consult a geological map of Europe will see at a glance that much of the continent was occupied by seas and lakes during the Tertiary period, and that it was not till towards the close of the epoch that as a land-mass it began to assume its existing configuration and dimensions. As with Europe, so to a great extent with Asia and Africa, but notably so with North and South America. The great land-masses had then been merely blocked out, as it were; their existing aspects are the results of Tertiary and Quaternary modifications.

sediments contemporaneously with the deltic deposits of the Po, Danube, and Volga, and these again contemporaneously with the silts of the Venetian and Hungarian lakes; so in tertiary times the same sort of operations went simultaneously forward, and thus we find throughout the system a variety not only of sediments, but of fossils, though all belonging to one great and continuous period of world-history.

There were also throughout the deposition of the tertiary strata abundant manifestations of volcanic activity, and in few tertiary districts are there wanting the cones, crateriform hills, lava overflows, ash-beds, and trachytic tufas that mark the comparative recentness of their production. In Central France, the Lower Rhine, Italy, Hungary, Greece, Western and Central Asia, Australia, and New Zealand, such evidences are everywhere abundant, and even in our own islands the basalts of Antrim and the trap-tuffs of Mull present their concurrent testimony. Less crystalline than the greenstones, felstones, and porphyries of the secondary and primary periods, and more compact than the lavas and cinder-beds of the present day, the tertiary traps are readily distinguished; and even where most consolidated, their age is easily determined by the associated strata. In Auvergne, the Lower Rhine, and Greece, they are found with beds of eocene and miocene age; in Australia, basaltic overflows cover auriferous gravels of pliocene date; and in New Zealand they spread over lignites of perhaps still later origin.

As a Life-period the tertiary system stands in remarkable contrast with all that we know of the preceding formations. In these the scheme of life, as now known to us, was incomplete—either some great order being largely predominant, or one or more orders altogether wanting. Taking the fauna alone, the Laurentian, Cumbrian, and Silurian were

characterised by the general absence of vertebrata; in the
Old Red Sandstone we had no reptiles, nor birds, nor mammals; in the Coal no birds nor mammals; and in the
secondary ages nothing apparently higher than marsupials.
But now, and for the first time in the history of the earth,
we are presented with all the great orders of plants and
animals, and were it not for certain forms that have become
extinct and others that are peculiar to the current epoch,
we could almost fancy we were dealing with the botany and
zoology of the present day. So far as they have been critically examined, the plants of the lower European tertiaries
indicate a much warmer climate than now prevails over the
same latitudes; and this higher temperature was in all likelihood brought about by the peculiar disposition of the sea
and land. Broad rivers flowing from tropical latitudes, and
inland seas extending longitudinally into sub-tropical or
even tropical zones, would be sufficient to account for the
presence of palms and other allied vegetation in the lignites
of Europe ; and what we already know of the boundaries
of the lower tertiaries, affords the best reason for believing
that such were the great geographical arrangements of the
period. In our reasonings on former climates we are too
apt to look to mere zones of latitude, without sufficiently
allowing for disposition and altitude of land, the trend of
oceanic currents, the prevailing set of winds, the effects of
atmospheric humidity, and other similar incidents by which
the luxuriance or sterility of life is often more immediately
affected than by mere propinquity to the equator. During
the deposition of the middle and upper tertiaries the climate
seems to have gradually declined, and the European lignites
of these periods exhibit a flora bearing a striking resemblance to that which now flourishes in North America.
We say " European lignites," for the lignites of Eastern
Asia, of New Zealand, and of British America seem each to

be characterised by its own peculiar plants, and these for the most part having a general relationship to the existing vegetation of the same regions. It must be admitted, however, that the flora of all the tertiaries—lower, middle, and upper —requires much more minute botanical investigation; and till this is done, geologists can offer little more than mere approximations to the conditions of the period.

Like the flora, the fauna of the lower tertiaries would seem to imply the existence of genial conditions at once of climate, food, and habitat. Gigantic sharks, turtles, crocodiles, and sea-serpents in the basins of London, Paris, and the Rhine, indicate much warmer waters than these latitudes now enjoy; while elephantine, tapir-like, camel-like, lion-like, and ape-like forms among their terrestrial fauna point unmistakably to sub-tropical or tropical surroundings. Nor is it mere variety of generic and specific forms among these mammalia, but their huge size and vast numerical abundance, that point in the same manner to favourable conditions of existence. Indeed, one of the most remarkable features in the life of the period is the vast bulk of the mammalia as compared with the same orders still existing. This massiveness of structure runs throughout all the divisions of the system—lower, middle, and upper—and marks alike the tertiary fauna of Europe, Asia, America, and Australia. If the number and magnitude of reptilian forms that thronged the secondary waters have conferred on that cycle the designation of "The Age of Reptiles," the number and magnitude of mammalian forms that peopled the tertiary lands may in like manner entitle this period to be signalised as "The Age of Mammals." *Palæotheres*, or tapir-like beasts, huge as elephants; *sivatheres*, or antelope-forms, tall as giraffes; *megatheres*, or sloths, weightier than the weightiest hippopotamus; *glyptodons*, or armadilloes, that could enclose a score of the living species under their shields; *macrauchenes*, or llama-forms, bulky as camels; *hyænodons*, or hyænas, stronger than tigers ; *diprotodons*, or kangaroos, heavier than oxen ; and *colossocheles*, or carapaced turtles, full fourteen feet in diameter, are but random instances of the colossal structures that have been exhumed from the sediments of the tertiary epoch. In this respect the fauna of the period seems to corroborate the idea that there is a culminating point in the life of orders, as there is in the life of individuals — a period when they attain their maximum development in numbers, variety, and magnitude, and after which they gradually decline and disappear, to make way for some newer and advancing order. Or as it has been put by one of the most recent writers on systematic geology (Professor Haughton), "it appears to be an almost universal law of life on the globe, that each group of organic beings increased in size and in importance in an uninterrupted line from the commencement of its existence, until its members reached their maximum in some short time—I mean, short as compared with their whole life-history—after their original creation and appearance upon the globe ; and it would almost seem as if, having reached that maximum of development, they then commenced a process of degeneracy and decline."

Another remarkable feature in the tertiary fauna is the prevalence of what are styled "intermediate forms," that is, of creatures partaking of the characteristics of two or more adjacent orders—a sort of interfusion, as it were, of families and genera, which now stand distinct and separate. We have thus elephant-like tapirs, camel-like stags, giraffe-like camels, horse-like antelopes, lion-like bears, tiger-like hyænas, and numerous other inosculating forms, which, had they existed now, would have filled up more closely the meshes of the great network of existence. Nor is it among the mammalia alone that this gigantic size and peculiarity of form make their appearance; for birds, reptiles, and fishes partake of similar characteristics, and point to the same favourable conditions of growth, and to the same great law of structural relationship and development. We say *structural relationship*, for, as has been well remarked by Professor Jukes, "in speaking of these extinct animals as forming links between our existing forms, we must never forget that the living forms are not the types, but the variations from the types. We are apt to assume that the forms with which we are most familiar are the most simple and natural; but the scientific naturalist often finds some extinct form as the simple archetype, from which numerous others have departed more and more by variation and combination of parts in subsequent periods." We are right when we speak of these tertiary mammals as holding an intermediate place between some existing forms, but we are wrong if we consider them on that account to be either more complex in structure or varied in function. "A three-toed horse (hippotherium) would now be looked on," says Mr Woodword, "as a *lusus naturæ;* but in truth, the ordinary one-toed horse of the present day is by far more wonderful."

We have already remarked, that though the distribution of the tertiary seas and lands differed considerably from the present, there must still have been a certain approximation to the present arrangement, inasmuch as the tertiary flora and fauna of every region, and especially of the later tertiaries, bear a considerable resemblance to the plants and animals that yet flourish there. There may be local differences among the tertiary basins of Europe and Asia; but still throughout the whole there is, if we may so phrase it, an aspect of Old World forms. The species and genera may differ, and there may be forms that stand intermediate

o

between two or three conterminous families; yet, on the whole, we see in the mastodons and mammoths, the palæotheres and anoplotheres, the hyopotami and chæropotami, the sivatheres and merycotheres, the machairodons and hyænodons, the prototypes and forerunners of the elephants, rhinoceroses, hippopotami, river-hogs, antelopes, giraffes, camels, horses, oxen, lions, tigers, and hyænas, that now inhabit the eastern hemisphere. In like manner the megatheres, mylodons, glyptodons, and macrauchenes of South America, were the gigantic forerunners of the sloths, armadilloes, and llamas that now people that continent; while the diprotodons, zygomatures, and nototheres of Australia, foreshadow the kangaroos and wombats so exclusively characteristic of that peculiar sub-continent.* Everything, indeed, both in flora and fauna, indicates the approach of existing nature; but this, as in all other cosmical operations, by slow and gradual steps—the eocene, miocene, and pliocene, each having its own special phases, and these diverging from those of the current epoch according to their relative distances in time.

What a picturesque and luxuriant scene these grassy plains and glades and river-banks of the old tertiary times must have presented! Uplands fretted with scrub, lagoons and river-creeks fringed with palms and tropical forest-growths, and swampy deltas teeming with tangling jungle. Herds of antelope- and horse-like forms (anoplotheres and hippotheres) scattered over the uplands; carnivora (hyænodon and cynodon) lurking among the rocks and bushes by the water-springs; elephants, tapirs, and wart-hogs (mastodons, palæotheres, and hyopotami) down by the

---

* By a perusal of the illustrations in any work on Palæontology, such as Owen's 'Fossil Mammals,' the non-scientific reader will be enabled to trace the resemblance between these extinct forms and the existing fauna far more readily than by any amount of verbal description.

river-side; and huge amphibia (halitheres, dinotheres, and crocodiles) lazily sunning themselves on the islets and mud-banks of the far-spreading delta ! Plain, lake, estuary, and sea, teeming alike with life during the warmer eocene; abounding in milder forms during the still genial miocene; and gradually assuming more temperate aspects as the pliocene approached more nearly to the ordainings of the current epoch. Such were the leading aspects of the Tertiary Times—times enjoying the genial surroundings of a peculiar geographical distribution of sea and land, and only brought to a close by new terraqueous arrangements inimical (over the greater portion of the northern hemisphere at least) to vegetable and animal luxuriance.

As Economic Repositories, the tertiary formations, though greatly inferior to some of the older systems, are not without their local importance. Brick and pottery clays, sands for glass-making, fuller's earth, tripoli or polishing-stone,[*] gypsum or plaster-of-Paris stone, limestones, burrh for millstones, and a great variety of lignites or brown-coals with their associated gums and amber,[†] are among their

---

[*] These polishing, infusorial, or microphytal earths are among the most wonderful of tertiary and recent accumulations, alike from their origin and the vast thickness to which they sometimes attain. According to Professor Dana, the infusorial earth of Virginia is in some places 30 feet thick, extends from Herring Bay in the Chesapeake, Maryland, to Petersburg, Virginia, or beyond, and is throughout an accumulation of the siliceous remains of microscopic organisms, mostly Diatoms. A still thicker bed, exceeding 50 feet, exists, according to Mr W. P. Blake, on the Pacific at Monterey, and is white and porous like chalk. The "polishing powder" of Tripoli and Bilin, and the "mountain meal" of Tuscany, Sweden, and other countries, are kindred accumulations.

[†] In some of the later lignites, like those of Northern Germany and Burmah, amber and other fossil gums and resins are found in all but their original attachment to the trees from which they were exuded. And in these ambers the insects of the tertiary forest are often as perfectly preserved as the specimens in the cabinet of the most fastidious collector.

most prevalent and valuable products.   These lignites, indeed, are year after year being discovered in new areas, and
are rapidly rising in economic value, as the only fuel in
many localities either for household or for manufacturing
purposes.   Germany, Italy, Austria, Hungary, Farther
India, New Zealand, Vancouver Island, the Northern
Prairies of America, and other regions, are each possessed
of their brown-coals, which will be more and more sought
after, and more and more valued, as manufactures, railways,
steamboats, and kindred modern appliances extend their
influence and demand the cheapest and readiest supply.
Compared with the older coals, many of the tertiary lignites
are no doubt inferior; but as the demand increases better
methods of consumption will be invented, and many of the
objections obviated which now stand in the way of their
being more largely raised both for domestic and manufacturing purposes.   There is thus no formation, however old
or however recent, that is not invested with industrial as
well as with scientific interest—all that is requisite being
research to determine the nature of their products, and skill
sufficient to procure and apply them.

ICE—ITS FORMS AND FUNCTIONS.

ICE, DEFINITION OF—GENERAL PHYSICAL PROPERTIES OF WATER—
FORMATION OF ICE—ITS OCCURRENCE IN THE ATMOSPHERE : HOAR-
FROST, SNOW, AND HAIL — ITS OCCURRENCE ON FRESH WATER :
RIVER, LAKE, AND GROUND ICE—ON SALT WATER : ICE-FIELDS,
ICE-PACKS, ICE-FLOES, ETC., IN POLAR SEAS—ICE ON LAND : SNOW
AND SNOW-LINE, AVALANCHES, NÉVÉ, GLACIERS—THEIR CHAR-
ACTERISTICS—ICEBERGS—GENERAL RESULTS OF ICE-ACTION.

THERE is no substance in nature so protean in form, so
incessant in circulation, or so multifarious in its functions
as water.   Now aëriform, now liquid, now solid ; now in
the ocean, now in the atmosphere, now percolating the
earth's crust, now coursing its surface and hurrying again to
the ocean ; now supplying the wants of plants and animals,
now wearing down the earth in one part, now accumulating
new material in another, and anon locked up, as water of
crystallisation, in the mineral structure of the globe for
ages.   The sum of its existence is change ; the whole course
of its history a series of marvels.   It is, however, only
with one feature of its existence, and with a small portion
of its history, that we have to do in the present instance ;
though we have necessarily to glance at its general deportment under heat, the incessant modifier of its form, and the
great propeller of its circulation.

At ordinary temperatures, as every one knows, water
appears in the liquid state ; at high temperatures it passes
rapidly into the vaporiform condition, and at low temper-

atures it becomes solid and crystalline. At 39° Fahr., or thereabouts, it appears to be at its maximum density; at that temperature and above it, under favourable conditions of the atmosphere, it is incessantly passing off as invisible vapour, and diffusing itself through the air; and at 212°, under the usual sea-level pressure of the atmosphere, it boils, and is rapidly converted into steam. From 39° down to 32°, it appears to suffer little change in density; but at and under 32° for fresh water, and 28½° for salt, it is suddenly converted into the crystalline or solid state, and is then known as *ice*. In this state it has expanded, become lighter, and necessarily floats on its own liquid surface. Ice, then, in ordinary language, is solid or frozen water. From its crystalline structure it occupies more space than when in a liquid state; hence its lightness and flotation, which are further increased by the number of air-cells which are always less or more entangled in its mass.*   Compared with water at 60°, whose specific gravity is 1, ice is found to be only .912; hence it floats with about one-ninth of its mass above water, and the remaining portion below. Being formed at 32°, ice may be said to be then in its normal condition; but at lower temperatures it slightly contracts, as was long ago proved by experiments on the ice of the Neva at St Petersburg.†   It is this ice, its various aspects and functions, that forms the subject of the present Sketch;

* It is only from water that has been subjected to boiling that ice free from these air-cells can be obtained.

† In further proof of this contraction, we may cite Sir James Ross, who says—"We have often in the arctic regions witnessed the astonishing effects of a sudden change of temperature during the winter season upon the ice of the fresh-water lakes. A fall of thirty or forty degrees of the thermometer immediately occasions large cracks, traversing the whole extent of the lake in every direction; some of the cracks opening in places several inches by the contraction of the upper surface in contact with the extreme cold air of the atmosphere." It is also partly by this contractile force that the ice-barriers and ice-walls of the polar seas are broken into floes and fragments.

its geological operations being those, of course, to which the attention of the reader will be more especially directed.

In treating of Ice in a popular way, it may be conveniently arranged under three great categories — ice in the atmosphere; ice on land; and ice on water. The ice or frozen water in the atmosphere is the great nursing parent of the ice on land; and the ice accumulated on the land becomes in the long-run one of the most remarkable features of the ice that floats on the water. As the vapour that ascends from the ocean is condensed and falls as rain on the land, and this rain finds its way again by runnel and river to the sea; so the vapour frozen in the air descends on the land, where, accumulating for ages, it slowly grinds and pushes its way once more to the ocean. In this way all the forms of ice are inseparably connected in one great and incessant round of circulation, and we only separate them provisionally for the purpose of intelligible description. First, then, as regards *the ice in the atmosphere*, the most casual observer must have noted the frequent formation of hoar-frost, snow, and hail. The rapid radiation of heat from the earth's surface, by which the invisible vapour of the atmosphere is converted into dew, has only to be carried beyond the limit of 32°, when *hoar-frost* is produced, crisping the herbage, or floating, in still conditions of the air, in clouds of crystalline spicules. A more rapid condensation of rain or vapour produces *hail*, which may occur at all seasons and under every latitude, and may fall in soft snowy drops, or pellets of ice, from the size of a coriander-seed to that of a pigeon's egg, and often with destructive effect on the crops of the farmer. Neither hail nor hoar-frost, however, exercises any perceptible influence on the rocky surface; and it is chiefly in the condition of *snow* that ice in the atmosphere becomes of interest to the geologist.

This snow, formed of frozen vapour, and composed of myriads of the most delicate geometrical crystals, occurs during
winter in all the higher latitudes, and at great elevations in
all latitudes, wherever the surrounding air falls below the
average of 32°.   As the atmosphere frequently consists of
strata of different temperatures, snow may be formed in the
higher regions, and yet in falling may pass through a warmer
stratum, and be melted into rain before it reaches the earth.
But in ordinary circumstances it falls on the land-surface in
soft downy flakes ; and if that surface be at or under 32°,
it often accumulates in great thickness, and especially in all
the higher and colder regions.   In the lower grounds it is
melted and carried off by the next thaw ; in the higher
mountains, where it falls in dry needle-like crystals,* and
rarely or ever in flakes, it may endure, summer and winter,
for generations.   But even there it cannot remain unchanged ; summer suns and the pressure of newer accumulations condense and urge it downwards—first as *névé*,† or
snow-ice, and ultimately as the pure transparent ice of the
glacier.   Here, however, it becomes ice on land, and falls to
be considered under a different section of our subject.

Like ice in the atmosphere, *ice on land* occurs chiefly in
the higher latitudes and at great elevations ; though under
a clear sky and extreme night-radiation, ice may be formed
on the ground even in sub-tropical and tropical countries.‡

---

* The sands of the burning desert are not so light nor so easily moved
as this dry crystalline snow-powder of the loftier mountains. The slightest
breath disturbs it ; the storm-wind sweeps it from the exposed heights,
and drifts it into the sheltered gorges in masses hundreds of feet in
thickness.

† *Névé*—the name given to the stratified slightly - compressed snow
of the higher Alps before it is condensed into the crystalline ice of the
glacier.

‡ The destructive effects of these night-frosts, under a clear, dry, and
serene sky, are now unfortunately too well known to our Australian sheep-

In our own islands, every one must have witnessed the effects of frost on the rocks and soils. The water held in the pores of all rocky substances is rapidly converted into ice; in this state it expands, pushes asunder the particles, and when thaw comes, the separated particles, having lost their cohesion, necessarily fall asunder, and are ready to be carried away by winds, rains, and runnels of water. The force with which water expands in freezing is tremendous. The strongest vessels are burst asunder; the hardest rocks are split into chips and fragments. Every winter we see its effects in the disintegration of our ploughed soils, and in the mounds of debris at the foot of our cliffs and precipices. But under our insular climate the effects are trifling compared with what takes place in higher latitudes and in more elevated regions. In the higher Himalayas Dr Hooker found the cliffs and precipices rent and rugged with its force, and the ravines choked with the ruptured blocks and fragments; in Norway every peasant can point to the mounds of angular blocks as the work of the " Bergrap;" and in Spitzbergen the Spanish Expedition found the sea-cliffs fresh with recent rupture—every winter severing with its icy wedge, and every summer dissolving the connection. It is needless, however, to multiply instances; every intelligent mind must perceive the power of this recurring ice-force, and its universality, in all the colder and higher regions of the globe. And he has only to allow sufficient

farmers. Even in the deserts of Africa, Arabia, and Persia, European travellers have felt their effects. In Bengal, where ice is never formed naturally, advantage is taken of the principle for its artificial production. Shallow pits are dug, which are partially filled with straw, and on the straw flat pans, containing water which has been boiled, are exposed to the clear, dry, and serene firmament. The water is a powerful radiant, and sends off its heat rapidly into space. The heat thus lost cannot be supplied from the earth, this source being cut off by the non-conducting straw; and before sunrise a cake of ice is formed in each vessel.

duration, and every peak and precipice would be rounded and worn down by its power.*

In all the higher latitudes, snow, we have said, falls less or more during winter, and melts away during summer.† But in all latitudes there is an elevation at which it lies perennially, and this elevation will differ, of course, with the latitude. This limit, above which snow lies at all seasons, is known as the *snow-line,* or *line of perpetual congelation;* and though it ascends a little higher during summer, and descends a little lower in winter, it is, on the whole, pretty stationary in every region of the globe. Of course, it will come nearer the sea-level in high latitudes, and ascend higher and higher as we approach the equator; and thus it is that we have it at 1500 feet in Spitzbergen, 2400 at North Cape, 5000 in the Dovrefelds, 9000 in the Alps, 12,000 in the Atlas range, and on an average about 16,000 feet under the equator. In all the higher regions, therefore, this snow accumulates enormously, and would continue to accumulate were it not for three causes which tend as incessantly to prevent it. These are, *first,* atmospheric causes, such as summer's heat, warm winds, and occasional rainfalls, which partially dissolve it; *second,* the mechanical pressure of the accumulating mass, which ever tends to urge it forward and downward to lower levels; and, *third,* the land-slopes, which afford greater or less facilities for its descent. As it descends by these means, so it melts

* The reader who takes interest in this matter will find marvellous illustrations of the power of frost in such works as Von Wrangell's ' Siberia,' Scoresby's ' Arctic Voyages,' Ross's ' Antarctic Voyages,' and Dr Hooker's ' Himalayan Journal.'

† Though snow is the necessary product of cold, yet in all temperate and coldly-temperate latitudes a good heavy snowfall is beneficial in protecting vegetation from the severity of long-continued frosts. Such a covering is usually known as the *snow-blanket,* and in central and northern Europe its absence in early spring is often followed by most destructive results to the young growths of the farmer and gardener.

away, and is carried by runnel and river to the ocean, again
to be raised as vapour, again to be frozen and fall as snow,
and again to be urged downward and melted to water.
Occasionally its descent from the mountains is sudden and
abrupt, as in the *avalanche*, which breaks away when the
gravity of the mass becomes too great for the slope on which
it rests, or when fresh weather destroys its adhesion to the
surface. These snow-slips, or rather snow-and-ice slips,
are frequently most destructive in their effects, and are the
dread of the traveller and inhabitant on mountain regions
like the higher Alps and Himalaya. These are usually
distinguished as *drift*, or those caused by the action of the
wind on the snow while loose and powdery; *rolling*, when
a detached piece of snow rolls down the steep, licks up the
snow over which it passes, and thus acquires bulk and impetus as it descends; *sliding*, when the mass loses its adhesion to the surface, and descends like a land-slip, carrying
everything before it unable to resist its pressure; and *glacial*,
when masses of frozen snow and ice are loosened by the
heat of summer, and precipitated with crushing effects into
the valleys below.

The great and persistent result, however, of this moun-
tain-snow, is the *glacier*—the " ice-sheet " of the flatter
heights, and the " ice-river " of the glens and ravines —
which is ever pushing and grinding its downward way till
it finally melts and becomes the gladdening stream of the
lower valleys. The snow that falls on the higher peaks,
being partially softened by the warmth and rains of summer, is converted into a sort of " snow-broth," or " slush,"
as the Scotch would call it, and has necessarily a tendency
to move downward by its own gravity, however gentle the
slope on which it rests. Pressed on summer after summer
by newer masses, it gradually assumes greater consistence,
loses the dull aspect of frozen snow, and passes into the hard

transparent state of the glacier or ice-river.　In its primary stage it is technically known as *névé* or ice-snow; and this névé, which stands intermediate between the pure unsunned snow of the winter heights and the moving glacier, is regarded as the fond or fountain of all true glaciers.　Ever fed by new snowfalls from above, it is gradually pushed downwards by the force of gravitation, and in turn propels the glacier, whose own weight and partial mobility also assist the downward movement.　The whole is, in fact, one great motion, just as it is part of the great circulation by which the water of the ocean is disseminated through the air and over the land, and the water of the land returned once more to the ocean.　Acquiring volume and weight as it descends (and some of the Alpine glaciers are from 80 to 600 feet in thickness), the glacier grinds and smoothes the rocks over which it passes; and this it does by the earth, gravel, and rock-debris which become incorporated with its mass, and which act like so many rasps and chisels on the rocky surface.　Slow in its motion, but persistent and irresistible, its course is ever downwards, and marked by abrasion, rounding, smoothing, and striation of the subjacent rocks.　And as it descends, the blocks and debris, loosened by the frost from the adjacent cliffs, fall on its surface, and are borne along in long winding spits, till the mass finally melts away in the lower valleys, and then this rock-debris is left in mounds or *moraines*.　These moraines—some of which are *lateral*, or on the sides ; some *medial*, or in the middle ; and some *terminal*, or at the melting end—bear ample testimony of the destruction that has taken place among the rocks above, and yet it is but a small portion of what has been discharged as impalpable mud by the discoloured stream that issues from the glacier.

Wherever glaciers occur — and they can only occur in mountains above the snow-line, and in regions where there is atmospheric moisture sufficient to produce continuous snows — their great geological function is to round and smooth down all rocky asperities, to round off projections, and produce *roches-moutonnées* as they are termed, to enlarge and deepen rock-basins, and to grind out and furrow the mountain glens down which they descend. And this descent, though obstructed by inequalities of surface, or retarded by the frosts of winter, never ceases. Cracked and crevassed as the ice-mass may be by the thaw of summer, hard and snow-clad as it is during the frosts of winter, it is still on the move, as persistent as gravitation itself, and as continuous as the snowfall by which it is created. This characteristic motion is thus summed up by Principal Forbes in his truly philosophical work ' On the Theory of Glaciers : '—1. That the downward motion of the ice from the mountains towards the valleys is a continuous and regular motion, going on day and night without starts or stops. 2. That it occurs in winter as well as in summer, though less in amount. 3. That it varies at all times with the temperature, being less in cold than in hot weather. 4. That rain and melting snow tend to accelerate the glacier motion. 5. That the *centre* of the glacier moves faster than the sides, as in the case of a river. 6. That the *surface* of the glacier moves faster than the bottom, also as in a river. 7. The glacier moves fastest (*other things being supposed alike*) on steep inclinations. 8. The motion of a glacier is not prevented, nor its continuity hindered, by contractions of the rocky channel in which it moves, nor by the inequalities of its bed. 9. The crevasses are for the most part formed anew annually, the old ones disappearing by the *collapse* of the ice during and after the hot season.

In the mountains of Norway, the Himalaya, and the Alps, the glacier winds its way down the glens and hollows till it reaches a certain limit in summer, and a little beyond that limit in winter; but still there is a limit at which it melts away and disappears, leaving its terminal mound of rounded blocks and shingly debris. In higher latitudes, however, such as Spitzbergen and Greenland, the *ice-sheet* that envelopes the land comes down to the sea-shore, and, ever urged forward by the formation of newer ice inland, even projects its icy wall far into the sea. But being lighter than water, there is a limit beyond which the mass, thick and heavy as it may be, cannot pass, and then it becomes buoyant, is broken off by storms, and drifted by winds and tides and currents as the *iceberg* over the surface of the deeper ocean. It is thus that the glacier, whether disappearing on the slopes of the mountain or melting away in the ocean, fulfils the beautiful saying of De Bouè, that " it begins in the clouds, is formed by the mountains, and ends in the ocean." These icebergs or ice-mountains are often of gigantic size, being several miles in circumference, and rising 50, 100, or 200 feet above the water.* And when it is borne in mind that

---

* The icebergs of the antarctic seas are generally larger, more precipitous, and more tabular in form than those of the arctic; while those of the latter, on the other hand, are more heavily laden with boulders, shingle, and land-worn debris. Sir James Ross thus adverts to some of the former :—" To-day (Jan. 31, 1841) several icebergs were seen ahead of us. They were chiefly of the tabular form, perfectly flat on the top, precipitous in every part, and from 150 to 200 feet high. They had evidently, at one time, formed part of the barrier (the great ice-barrier that prevented his approach to the southern pole, and which was estimated at more than 1000 feet in thickness), and I felt convinced, from finding them at this season so near the point of their formations, that they were resting on the ground. The lines were immediately prepared, and when we got amongst them next morning we hove-to, and obtained soundings in 1560 feet, on a bottom of stiff green mud, leaving no doubt on our minds that all the bergs about us, after having broken away from the barrier, had grounded in this curious bank, which, being two hundred miles from Cape Crozier, the nearest known land, and about sixty from the edge of the barrier, was of itself a discovery of considerable interest."

little more than an eighth of the mass makes its appearance above the surface, one can readily form some conception of the bulk and weight of a " berg," and of its crushing and grinding power when drifted along at the rate of four or five miles an hour. But the ice that was generated on and has now become ice on water, and belongs to the next division of our subject.

But before taking leave of ice as it appears on the land, there is another condition in which it not unfrequently occurs, and which is well deserving of the attention of the geologist and hydrographer. We allude to its formation and conservation in caverns and fissures where it may have lain unchanged for ages. These *ice-caves*, as they are called, are found in many countries, and most abundantly, of course, in all high, dry, and cold regions. In some the ice appears in stalactitic and stalagmitic masses, issuing from the roofs in fantastic cascades and wall-like screens, or resting on the floor in cones and pillars ; in some it merely forms a pavement of varying thickness ; while in others it glitters on all sides like a casing of the purest alabaster. Palæontologically, the preservative effects of such cavern ice must be very great ; hydrographically, its partial melting in summer may feed the mountain-springs that would otherwise be dry ; and economically, it yields to the surrounding districts its cooling and refreshing supplies. In Europe the better-known ice-caves or *glacières*, as they are locally termed, occur in France and Switzerland, and those form the subject of a recent volume by the Rev. G. F. Browne, a perusal of which will well repay the reader.* As the subject is somewhat novel, the theory of their formation, as given by Deluc and the author, may prove instructive ; and this we present with a little condensation :—

" The heavy cold air of winter," says Mr Browne, "sinks

* 'Ice-Caves of France and Switzerland.' Longmans & Co. 1865.

down into the glacières, and the lighter warm air of summer cannot on ordinary principles of gravitation dislodge it, so that heat is very slowly spread in the caves; and even when some amount of heat does reach the ice, the latter melts but slowly, for ice absorbs 60° C. of heat in melting; and thus when ice is once formed, it becomes a material guarantee for the permanence of cold in the cave. For this explanation to hold good it is necessary that the level at which the ice is found should be below the level of the entrance to the cave; otherwise the mere weight of the cold air would cause it to leave its prison as soon as the spring warmth arrived. In every single case that has come under my observation this condition has been emphatically fulfilled. It is necessary also that the cave should be protected from direct radiation, as the gravitation of cold air has nothing to do with resistance to that powerful means of introducing heat. A third and very necessary condition is, that the wind should not be allowed access to the cave; for if it were, it would infallibly bring in heated air in spite of the specific weight of the cold air stored within. There can be no doubt, too, that the large surfaces which are available for evaporation have much to do with maintaining a somewhat lower temperature of the place where the air occurs. Another great advantage which some glacières possess must be borne in mind—namely, the collection of snow at the bottom of the pit in which the entrance lies. This snow absorbs in the course of melting all which strikes down by radiation, or is drawn down by accidental turns of the wind; and the snow-water thus forced into the cave will at any rate not seriously injure the ice." So much for ice as it appears on the land; let us next turn to its aspects and functions on the water.

The last and most obvious division of our subject, perhaps, is that of *ice on water*. In all the temperate and colder latitudes this is a common winter phenomenon—occurring whenever the thermometer falls below 32°, and most rapidly in clear, dry, and serene conditions of the atmosphere. Every one who has watched by the stagnant pool must have observed the first formation of ice—a creeping or shrinking, as it were, of the surface occasioned by the incipient ice-crystals, which shoot hither and thither, interlace, and coalesce till a thin continuous crust has gathered over the whole. This is the first film; and as freezing takes place from above,* every successive film is formed more slowly; the ice-crust as it thickens protecting more and more from the cold the water that lies below. Indeed it is often curious to observe how little will obstruct the radiation of heat and prevent the formation of ice. An overhanging tree, a few leaves drifted over the first film, or even a cloudy sky, is sufficient to retard or obstruct; and though a clear and serene sky be in general most favourable, yet an air of dry wind to remove any superincumbent vapour will materially promote the operation. As the frost continues the ice thickens, but not indefinitely, for in water of sufficient depth this thickening acts as a barrier to its further increase, and even in the coldest regions it is only the shallower waters that are ever completely converted into ice.

Besides the ice that forms on the surface of fresh water,

---

* This principle of freezing from above is not sufficiently taken advantage of by our skaters, curlers, and ice-storers. Instead of waiting till the frost has produced a sufficient thickness of ice in the natural way—a thing never to be depended on in our uncertain climate—they ought to have the water let by degrees over the surfaces of the already formed ice, and in this way its thickness and strength would be rapidly augmented. We believe the celebrated ice of Lake Wenham, now so largely used in and exported from North America, is treated in this way—each successive surface being scraped and cleaned before the next overflow of water. For hardness, transparency, and general beauty of mass, the Wenham ice is unrivalled.

there is occasionally witnessed the rarer phenomenon of *ground-ice*, or that which gathers in thin sheets along the pebbly beds of shallow lakes and streams. These pebbles, losing their heat by radiation quicker than the water—for heat radiates through as it does from the surface of the transparent water—act as points for the formation of ice-crystals, and these passing from stone to stone, shortly convert the entire pebbly bed into a crust of ice. We have seen in a shallow ford of a Scottish stream ground-ice fully an inch in thickness when the stream itself was still flowing and unfrozen. This ground-ice, when broken up by freshets or other causes, floats down the stream, bearing with it its burden of encrusted pebbles, and thus becomes in nature a curious means of geological transport. Pebbly gravels may thus be laid down in situations where no current of water could carry them ; just as Deas and Simpson found the ice-cake of the arctic shores driven forty or fifty feet above the sea-level, and as it melted away, leaving long ridges of beach-gravel at heights to which no ordinary wind-wave could ever transport it.

But important as the formation of ice on fresh-water lakes and streams may be, it is as nothing compared with the masses that accumulate on the surface of the arctic and antarctic seas. Instead of forming at 32°, ice does not appear on salt water till the temperature has sunk to 28½° ; and then it goes on increasing, according to Sir Edward Belcher,* at the rate of about half an inch per day, during the long polar winters, often attaining a thickness of 10, 15, and 20 feet. This enormous crust, as it stretches unbroken over the ocean, is the

---

* Sir Edward's observations were made in Wellington Channel (1852–54) when in search of the missing Franklin Expedition. See his Narrative for some curious and instructive facts respecting the formation, character, and deportment of polar ice.

" ice-field " of the sailor, which, when broken up, becomes his "ice-brash," and either floats away in "floes" and "patches," or is drifted by winds and currents in " packs " and " streams." * The conservative effect of this ice-crust on the warmth of the ocean is one of the most providential arrangements in nature—maintaining for the water beneath the mean of 39°, when in the air above the cold is often sufficiently intense to solidify mercury, whose freezing point is—39°, or 39° below zero. Gradually as the ice thickens, it protects more and more the subjacent water; † gradually as the water in contact with the under surface of the ice is chilled, it becomes heavier, and sinks, its place being taken by a warmer film; and gradually as the water is converted into ice (it freezes fresh, or only with such brine as may be entangled in its interstices), the upper film, being salter and denser, descends, and lighter and warmer particles ascend to take its place. In freezing, water, of course, gives out heat, and the heated air-bubbles may often be seen clustering beneath the ice and struggling as it were to escape upwards. Every bubble in this way

---

* The names by which the different conditions of sea-ice are known to our whalers and navigators. The " ice-field " or " field of ice " is the unbroken ice of the polar oceans ; when broken up by thaws and storms it becomes "brash-ice ; " when drifted into dense masses it is "pack-ice ; " and when floated away by winds and currents it passes either into solitary " floes," into "patches" of several floes, or into " streams," having a determinate direction. A solitary fragment floating with a considerable portion of its bulk above water is a " hummock ; " and when loaded with debris and chiefly under water it is a " calf." The young ice that is rapidly formed, on the approach of winter, between floes and patches is "pancake ice ; " when of greater thickness, and formed in creeks and inlets, it is "bay-ice." These different conditions are also known at a distance by their " blink " or reflection—this being *clear* for field-ice, *white* for packed, *grey* for newly formed, and *deep yellow* for snow.

† As ice slightly contracts at temperatures under 32°, the intense cold of the polar regions only tends to render it more homogeneous and compact, and thus to increase still further its powers of protection.

melts its modicum of ice, and one by one, as they ascend in
the same direction, they gradually pierce the thickest sheet,
cutting rounded holes as clean and straight (in the words of
Sir E. Belcher) as if they had been bored by an auger.   In
general, however, the heat is retained and diffused throughout the water, while all above is stark and lifeless at temperatures 50°, 60°, and even 70° below zero.   How wonderful the provision by which the density and temperature of
the ocean are preserved for the wants of its animal life !
how perfect the scheme of compensation by which the most
powerful agents are held in check, and the balance and
equipoise of nature sustained !

It is at this stage, when the thaws and currents of a brief
summer have broken up the polar ice into " floes " and
" packs " and " streams," that we find it associated with the
land-formed " berg ;" the whole drifting to warmer latitudes, there to be dissolved, and to lose themselves once
more in the liquid mass of the ocean.   Purely sea-formed
ice has no perceptible geological effect, but much of it is
accumulated along shore and under cliffs and precipices
(the " ice-foot " of the sailor), and this, along with the true
iceberg, is generally laden with soil, sand, gravel, bouldery
blocks, and other spoils of the land, and these, as the ice-
masses melt away, are dropped broadcast over the floor of
the ocean.   All that is, or has ever been, ground and worn
from the surface of Greenland by the ice-sheet that envelops
it, has been spread by the iceberg on the bottom of the
North Atlantic.   Water in the solid state is as much a
wearer and transporter of the land as water in a liquid state.
The ice-stream grinds and degrades as surely as the water-
stream ; and the burden of both ultimately finds its way to
the depths of the ocean.   Nay, ice is the more potent of
the two—the " berg " bearing blocks and boulders which
no current of water could move, and scattering its burden

over the outer depths of the ocean, while the river-sediment is merely fringing the inner shores.

Such is a brief review of the various aspects in which ice occurs, and the more prominent functions it appears to perform in the economy of nature. As snow in the lower grounds, it acts as a protecting *blanket* against the severity of long-continued frosts ; as snow in the higher regions, it passes into *névé* and *glacier* to grind and round the rocky surface in its descent, and to smooth into gentler outlines the asperities over which it passes in its slow but irresistible progress. As the liquid stream erodes and deepens its channel, so the ice-stream rasps and · chisels—the function of both being to wear and degrade the old rocks, and to transport the material for the formation of the new. As ice on water, its greater bulk, as compared with that of ,the water from which it is formed, enables it to float as a protecting surface, preventing the water below from being entirely frozen, and thus preserving a habitable medium, no matter how intense the cold, or however long it may be continued. As ice on water also (the iceberg) it becomes a geological carrier, transporting to the outer depths of the ocean the gravel and shingle and boulders of the rocky shores, and piling them up in long submarine reaches according to the set of the tides and currents by which they are mainly directed. As ice in the rocks and soils, it is ever splitting and disintegrating ; unless within the limits of perpetually frozen ground, as in the tundras of Siberia and the swamps of Arctic America, and there it exercises a conservative effect *—binding the softest soils as hard as rocks,

* Even within these icy flats the power of frost is sometimes curiously destructive. "The influence of the cold," says Von Wrangell, speaking of the December temperature of Siberia, which was 58° below freezing, "extends even to inanimate nature. The thickest trunks of trees are rent asunder with a loud sound, which in those deserts falls on the ear

and preserving their imbedded organisms fresh and unchanged for ages.*   In all its aspects, ice is invested with
a curious interest ; in all its functions it is charged with
important results.   To us, the inhabitants of an insular
and unstable climate, it may appear of little importance ;
but to those of the higher latitudes and altitudes it assumes
the boldest character, and achieves the most gigantic results.   And these results, when accumulated for years and
ages, present to the geologist, as we shall see in the following Sketch, phenomena as marvellous in magnitude, and as
complicated in character, as those produced by any other
agency to which the crust of our earth is subjected.

like a signal-shot at sea ; large masses of rock are torn from their ancient
sites ; the ground in the tundras and in the rocky valleys cracks and forms
wide yawning fissures, from which the waters that were beneath the surface rise, giving off a cloud of vapour, and become immediately changed
into ice."

* It is chiefly in the frozen sands and gravels of the Siberian lowlands
that the remains of the mammoth and woolly rhinoceros are preserved
·in greatest perfection.   Every geological reader is acquainted with the
history of the St Petersburg specimen ; how hair, wool, and muscle were
so fresh, when first discovered, that even the dogs of the Tungusian
hunters were tempted to feed upon them, and this after the entombment
of ages !   The manner of their occurrence is thus described by the autho-
rity above quoted : " The banks of the rivers consist of sand-hills 150 or
200 feet high, and held together by the perpetual frosts which the summer is too short to dissolve.   Most of these hills are frozen as hard as
rock ; nothing thaws but a thin outside layer, which, being gradually
undermined by the water, often causes large masses of frozen sand to
break off and fall into the stream.   When this happens, mammoth remains, in more or less good preservation, are usually discovered."

THE GLACIAL OR ICE EPOCH.

THERE are few chapters in geological history possessed of
more interest than that which is usually known as the Glacial Epoch, and none that has received a larger share of
attention from modern investigators. Indeed, it is still to a
great extent an unsolved problem, and hence the conflicting
views that prevail as to the physical conditions of the period,
and the causes by which these conditions were produced.
Though beset with many difficulties, the general features of
the period are well known, and it is to place these broadly
before the general reader, rather than enter upon debatable
hypotheses, that we attempt the present Sketch.

As mentioned in a preceding paper, the genial temperature that prevailed during the deposition of the earlier tertiaries began gradually to decline during the middle and
later portions, till towards the close of the period an intense
cold set in, and ice seems to have prevailed alike over the
land and waters. Of course, we refer more especially to the

higher latitudes of Europe, Asia, and North America, the
regions within which the phenomena of the glacial epoch
are most strikingly displayed, and to which (from the 40th
or 42d parallel northwards) they were in all likelihood restricted.   Over these limits the ice-epoch long held its iron
sway, annihilating, or all but annihilating, terrestrial life ;
grinding, and rounding, and moulding the land-surface as
no other agent but ice can do ; and loading the bottom of
the ocean with miscellaneous masses of mud, shingle, and
boulders.   This paucity of life, these land-surfaces, and
these miscellaneous accumulations, are the principal proofs
of the conditions of the glacial epoch, and these we must
first consider under the knowledge we have gained by a
perusal of the preceding chapter.

At first it seems evident that towards the close of the
tertiary period the climate of a large portion of the northern
hemisphere was gradually growing colder and colder.   In
all likelihood the land was somewhat higher than it is
now, and as this cold increased the loftier mountains would
become perennially enveloped with snow and glacier, and
the surrounding seas with an annual covering of ice.   Under this increasing rigour all the more delicate tertiary
plants and animals would succumb, and those endowed
with greater elasticity of constitution would shift ground
to lower and more southern situations.   As the cold still
increased, the ice-sheet seems to have spread itself even over
the lower grounds, to have pushed its way out to sea, and
during the thaw and currents of a brief summer to have
been drifted off in floes and bergs, as the ice is now from
the coasts of the arctic and antarctic regions.   At this stage
the terrestrial flora and fauna would be at their minimum,
and paralleled, perhaps, by what we now find in Greenland
and the islands of the Arctic Ocean.   During this setting-in

of the glacial epoch, the land, as we shall shortly see, seems to have been gradually subsiding, and this subsidence went on to the extent of 1800 or 2000 feet below the existing sea-level, converting a large portion of what is now Europe and America into series of frozen straits and ice-clad islands. When the land seems to have been at its greatest depression the cold appears to have attained its greatest intensity, and at this stage we have the zenith and turning-point of the glacial period. After the lapse, perhaps, of ages, a reverse action sets in ; the land begins to be re-elevated ; a new cycle of temperature commences ; and the cold, though still clinging in snow and glacier to the higher hills, is less felt along the lower grounds and neighbouring sea-shores. By-and-by, as the elevation continues, the glaciers melt away from the hill-sides ; the icebergs and ice-packs disappear from the seas ; the general climate improves ; plants and land animals in newer species gradually take possession of the land ; and the existing order of things is imperceptibly established. Such seem to have been the setting-in, the creeping-on, the culmination, and the departure of the glacial epoch. Let us now glance at the proofs by which this advent, this subsidence, and this re-elevation can be logically established.

That the ice-epoch, like other great events in nature, came on slowly and gradually, is abundantly evidenced by the temperate or even coldly-temperate aspects of the flora and fauna of the later, as compared with those of the middle and earlier tertiaries. The eocene palms, crocodiles, turtles, and monkeys do not appear in miocene strata ; the miocene sycamores, chestnuts, and maples are replaced by pliocene pines, beeches, and birches ; and thus over the tertiary areas of Europe at least the declension of climate had been going on for ages before the advent of the glacial period. How far this declension was simultaneous over Asia and America has not been determined, but that a similar declension took place in those areas is sufficiently obvious from a similar change in their flora and fauna.*   That the pre-glacial land was somewhat higher than the present is shown by old river-courses and land-surfaces that lie below the existing sea-level, as well as by ice-marked rocks that dip away beneath the waters.   Had the pre-glacial lands been lower, these rock-surfaces would not have been smoothed and furrowed by ice, nor would the old land-surfaces have made their appearance.†   It was on this more elevated surface, therefore, that the glacier and ice-sheet first began their operations ; and it is at this stage that we find the lowest tenacious clays ("lower till"), and angular blocks and boulders, little removed from the rocks from which they were severed.   Here, then, we have the *first stage* of the glacial epoch—the operation of ice on a land-surface somewhat more elevated, in its average altitude at least, than the existing continents of Europe, Asia, and North America. This operation, as we have learned from the preceding Sketch, must have been to grind and gouge in the valleys, to smooth and round the higher eminences, and generally to polish the harder rocks—the detritus or abraded material being carried forward to lower levels, there to be laid

* According to Professor Dana, there are no tertiary strata in North America to the north of the New England States, the northern area having been dry land while the southern was under water and received the tertiary deposits.   To this elevation of the northern lands, and the subsequent gradual uprise of the southern or tertiary portion to the height of 3000 or 5000 feet, he attributes the first setting-in of the glacial epoch.

† The attention of geologists has not been sufficiently directed to these pre-glacial land-surfaces.   It is true that the "lower till" rests for the most part on abraded rock-surfaces, but there are many localities (we have noticed them in Kincardine, Ayr, Fife, and Durham—to say nothing of the well-known Cromer beds in Norfolk) where it reposes on sands, gravels, and even peaty beds, which were undoubtedly the soils and river-gravels of the period immediately preceding ; and in these we may expect to find the remains of the true pre-glacial flora and fauna.

down as clay and mud, mounds of shingly gravel, and masses of blocks and boulders. Whatever the nature of the parent rock, these mounds and masses would partake of it—yellow clays and schists and granites in granite districts, red clays and red sandstone blocks in Old Red Sandstone areas, and dark-coloured clays and blocks of limestone and sandstone in Carboniferous basins; and as a general rule these blocks and boulders not far transported from the cliffs and precipices from which they had been torn by the ice-giant. Indeed, in most instances, this proximity of glacial clays, ice-worn and ice-scratched blocks, is one of the best proofs of the first stage of the ice-epoch, and all over the northern and middle portions of Scotland we have never found it to fail in its indications.

But as the cold set in more intensely, the downward movement of the land seems to have commenced; and hence much of the ice-worn debris of this first stage was removed by denudation, step by step, as the terrestrial surface descended. What may have been the precise character of the climate at this epoch of descent—that is, how long the winter frosts and how short the summer's thaw—we have no means of determining, for the earlier clays and moraine blocks are destitute of organic remains; but if we may judge by the comparatively small amount of obliteration by denudation, it would appear that the seas were more ice-locked than free-flowing, and that consequently the land went down encased, as it were, in ice of prodigious thickness.* This descent or subsidence of the land forms the

---

* The ice-sheet at this stage may have been two or three thousand feet in thickness. The great antarctic ice-barrier, met by Sir James Ross and his companions, was estimated at a thousand or fifteen hundred feet. Ice of this thickness would rest on the beds of all the shallower friths and seas, and act upon them precisely as upon the rocky surfaces of the dry land. This circumstance should be carefully borne in mind in reasoning on the phenomena of the glacial epoch.

second stage of the glacial epoch, and must have been characterised by all that glacier on land or iceberg on water are capable of performing.   The ice-sheet that now gathered over the gradually-decreasing land would push its way shoreward with its annual burden of debris, and this debris, as it was carried seaward, would be dropt in part over the clay, shingle, and boulders that had been accumulated during the first stage.   Of course, a considerable portion of the debris of the first stage would be removed by shore denudation as the land subsided, but a large portion was also left undisturbed ; and thus it is that we find in many places the earlier clays and angular blocks covered over by other clays, replete with boulders more worn and rounded, and more strongly marked by scratches and furrows.   It was then, and during this period of subsidence, that huge boulders were carried by floating ice far from their parent rocks; and thus it is that these boulders, now hundreds of miles from their original cliffs, mark in a special manner the second stage of the glacial epoch.*   During this stage, as during the preceding, we have no evidence of a terrestrial flora or fauna, the climate being evidently too rigorous for their support; and it is only towards the southern limits of the ice-field (the 40th or 42d parallel of latitude †) that we can expect to find the remains either of

---

* It is difficult to convey by description the difference between the clays and boulders of the first and second stages ; but a few days in the field will train the eye sufficiently to mark the distinction, and this altogether independent of superposition.   There is a roughness of admixture and heterogeneousness about them that never appear in those of the earlier stage.   Perhaps the best test of the second stage is the number of " erratic blocks," or boulders far removed from their parent rocks.   In Europe, Scandinavian rocks are found in Central Germany and over the south-east of England; in North America, Canadian blocks occur a hundred and a hundred and fifty miles southward of their parent sources.

† American geologists give the southern limit of the drift in their

plants or animals, and these in all likelihood of boreal habits, like those now inhabiting the borders of the arctic regions. How long this period of descent continued relatively to the other stages, it is difficult to determine, though, on the whole, it appears to have been the longest stage, and that which has most impressed its character on the terrestrial surface. The extent to which the subsidence took place is variously estimated at from 1700 to 2000 feet, for from that limit down to the existing seashore, the land-surface is marked with rounding and smoothing, polishing and scratching, glacial moraines and clays, ice-borne blocks and boulders. One cannot turn to the higher districts of our own islands, to the north of Europe, or to Northern America, without perceiving on every hand traces of this long-continued ice-action—the bouldery clays, the rounded blocks and boulders, the scratched and polished rock-surfaces, the rounded outlines of the hills and knolls, all bespeaking its presence as incontestably as the existing surface of the Alps, the Scandinavian mountains, or the uplands and shores of Greenland.

But the forces that govern the external conditions of our planet are never at rest. Change succeeds change, and cycle follows cycle. The downward tendency of the land ceases, and an upward movement commences. Along with this gradual elevation, new distributions of sea and land begin to appear, and with these changes the intensity of the glacial epoch seems to come to a close. Glacier and ice-sheet, however, still shroud the land, and icebergs drift away from the shores. Other currents, however, are evidently setting in, a more genial climate begins to prevail, and with this higher temperature the ice disappears frum the seas and lower grounds, and only clings to the higher hills in

continent as 39° N. lat. ; in Europe it has been variously stated at 40°, 42°, and even 44° N. lat.

shrinking and gradually lessening glaciers. Even these, too, vanish in the long-run, and the present order of things, the ordainings of the current epoch, are established. This gradual elevation of the land constitutes the *third and last stage* of the glacial epoch, the proofs of which are to be found in the moraines, lateral and terminal, that still linger in every glen and corrie of our island, in the re-assorted clays and boulders of the two former stages, in the numerous terraces which mark the successive steps of the land's uprise, and in the fine silty clays (the " brick-clays" of some geologists), with boreal shells, star-fishes, bones of seals and whales, which fringe our bays and estuaries at various altitudes above the present sea-level. Compared with the earlier stages of the ice-epoch, the traces of this latter stage are still fresh and recent. The mounds of sand and gravel so frequent at the mouths of all our glens and upland valleys, are but the terminal moraines of the ancient glaciers ; the gravelly terraces (in some instances with shells) that fringe so many of our hill-sides, are but the ancient beaches of the gradually uprising land; the great blocks and boulders so abundantly strewn over our heathy uplands, are but the denuded ice-borne blocks of the two former stages ; and the fine silty " brick-clays" are but the upraised muds of the deeper sea-bed. Since then, frost, rain, and rivers have done their work on the land's surface, and obliterated many of the ice-traces, yet enough still remains to convince the unbiassed inquirer of its long reign over these northern latitudes during the three successive stages we have here endeavoured to describe.

The accumulations described in the preceding paragraphs are usually distinguished by such names as " northern drift," " glacial drift," " erratic blocks," and " boulder-clay," all conveying the idea that they have not been deposited under the ordinary conditions of water, that they belong especially to the northern hemisphere, have been drifted from a northerly source, and in all likelihood by ice partly on land and partly in water.* The older terms "diluvium" and "diluvial drift," under the idea that they were the results of the Noachian deluge, have been long ago set aside, and geologists with one consent now look to ice, in one or other of its manifestations, as the only known agent by which they could have been accumulated. Rain and rivers can, no doubt, waste and wear down the land, but it is ice alone that can grind and smooth and confer those rounded outlines which characterise so much of the surface in the higher latitudes of the northern hemisphere. River floods and freshets can accumulate vast mounds of sand and shingle; but it is the glacier alone that throws across the glen its dam-like moraine, and scores and polishes the pebbles of which it is composed. Running water laden with debris will wear and smooth the rock-surfaces over which it flows; but it is ice alone that can put on the glassy polish, and scratch and gouge in long straight furrows with its imbedded blocks and fragments. Rivers and torrents will roll and transport blocks of considerable size; but it is ice alone, either as the ice-sheet on land or the iceberg on water, that can tansport boulders many tons in weight up and over hills, or float them away hundreds of miles from their parent precipices. Running water invariably assorts its debris in beds and layers according to its fineness; it is floating ice alone that drops its burden on the sea-bed without regard to arrangement. All these appearances—boulder-

---

* Objection has been taken to each and all of these designations, and it must be confessed we are still in want of a good general term for the accumulations of the glacial epoch. Drifts, erratic blocks, and boulder-clays, are but members of a great series, and it is for this series, taken as a whole, that we still stand in need of a comprehensive designation.

clays of great thickness, erratic blocks of enormous size, polished, striated, and grooved rocks, moraine-like mounds of gravel, and smoothed and rounded surfaces—are so common in Britain, Northern Europe, and North America, that geologists are driven to the conclusion of a glacial epoch; a long period intermediate between the Tertiary and the Current era, when all the northern hemisphere, down to the 40th or 42d parallel, was under the influences of an icy climate like that which now prevails within the arctic and antarctic circles.

Geologists have long been at variance, and in some instances are still at variance, as to whether the phenomena are to be attributed more to land-ice or to sea-ice, to glaciers or to icebergs. As we became better acquainted, however, with the operations of ice in such regions as the Alps, Himalayas, Scandinavia, Spitzbergen, Greenland, and the polar seas, such differences of opinion grew less, and competent authorities seem to be agreed that we must call in both agencies, and this during the successive stages of subsidence and re-elevation which we have already endeavoured to describe. Indeed, it is impossible to conceive of a glacial climate over any large portion of the earth's surface without seeing that it must affect sea and land alike; and that if there is any brief period of thaw, like the existing arctic summer, the ice must be set in motion both on land and water.* Once set in motion, each would contribute its quota to the general result—the land-ice grinding and smoothing and rounding the rocky surface in its descent to the sea, and the sea-ice ploughing the shallower sea-bed as it floated away to drop its burden of boulders and debris

---

* Even without any great degree of summer thaw, a mobile mass like ice and snow would be urged forward by its own accumulating weight, and this over heights and hollows, so long as the head pressure remained sufficiently powerful. See preceding Sketch, p. 221.

in the deeper and warmer waters. In consequence of the long subsidence and gradual re-elevation of the land, the results of the terrestrial and sea-floated ice would become commingled; and hence one great source of the difficulty attending the interpretation of the phenomena of the glacial epoch. He, therefore, that would read aright the ice-epoch, must restrict himself to no limited theory, but must grasp broadly the consequences of a general glacialised condition, calling in alike the aid of terrestrial and marine ice, and ascribing to both what he cannot clearly determine as having been performed by either.

Another question that has given rise to much discussion, is the direction in which the ice has generally moved, and the source or sources from which the debris has been borne. In the north of Europe the set seems to have been from the north or a little west of north, so that the rocks torn from the Scandinavian mountains have been scattered over the plains of Denmark and Germany. In the British Islands, the set seems also to have been from the north and north-west, so that boulders from the Grampians are strewn over the Lowlands, and blocks from the Cumberland hills over the valleys of Yorkshire. In North America too, the general float or trend has been, in like manner, from the north, or from directions between the north and north-west—hence the appropriateness of the term "northern drift" when applied to the erratic blocks that now lie scattered so far and wide from their native sources. But in most countries the land-ice has moved in accordance with the general slopes of the land or the trends of its principal valleys; hence the glacial gougings and groovings radiate in all directions from some main eminence, or lie chiefly in one way, according to the bearing of the valley down which the ice has descended. A great deal of minute but worthless labour has indeed been spent on this subject. The grooves

Q

and striæ produced by the mountain-glacier can give no idea of the direction of the current that bore the iceberg; and the ploughings of the iceberg through firth and strait can give no indication of the course of the glacier from which it was detached. The fact is, that the direction of these groovings and gougings must be read each in the light of its own locality; and we have seen on many rock-surfaces in Scotland striæ crossing each other, and evidently produced at different times by forces moving in different directions. That the land-ice must have moved in the direction of the slopes on which it rested, is self-evident; and that the floating ice must have taken a southerly course, is only what may have been expected, seeing that the cold currents from the poles must ever be towards the warmer and lighter waters of the equator.

A third and important question is, whether this glaciation of so large a portion of the northern hemisphere was contemporaneous, or whether it passed gradually and successively over the areas of Europe, Asia, and America? This question, in fact, involves the cause of the glaciation, and will be answered very much according to the views which different theorists entertain. If we regard the ice-epoch as brought about by external or astronomical causes, its contemporaneity is the most likely result; but if we consider it as arising from some peculiar distribution of sea and land, and dependent on the earth's own physical relations, the glaciation of Europe may have been separated by long ages from that of America; or, in other words, the ice-mantle may have gradually crept from the one hemisphere to the other, according as the set of polar currents was altered by upheaval and depression of the land areas. In the mean time, the general leaning seems to be towards the idea of contemporaneity, though it must be confessed that a gradual advance of the glaciation from area to area seems, in some respects, the more philosophical belief. Either way the fact remains of its occurrence between the tertiary and current epochs; of its having brought to a close the tertiary conditions that prevailed over these latitudes; and of its having passed away before the present order of things could be established. Wherever it occurs it forms a great hiatus between the Tertiary and Recent—a lifeless blank, as it were, between the flora and fauna of the Tertiary epoch, and the flora and fauna now flourishing around us.

The hypotheses that have been advanced to account for the glacial epoch resolve themselves into two great categories—*first*, those that depend on astronomical causes, or causes extraneous to the earth; and, *second*, those that depend on the earth's own physical relations as regards the then peculiar distribution of its lands and waters. This is not the place to enter minutely into the merit of these theories, but we may glance at their general bearings as indicating the widely different sources to which geologists have been driven for a solution of this strange and anomalous period. By those who seek for the solution in a peculiar distribution of sea and land, it is argued, that by elevating the terrestrial surface in the northern hemisphere and shutting off the warm currents of the ocean from the circumpolar areas, an extreme degree of cold would be induced; that the snow and ice accumulated under this cold would have a tendency to diminish still further the annual temperature; and that by these means the ice-sheet, in course of ages, would push its way southward alike over land and sea, to the furthest known verge of glaciation. This, they contend, would account for the first stage of the ice-epoch : and even when, during the second stage, the land began in some areas to subside, the ice-sheet must have acquired so much mastery over the average annual temperature as still to maintain its ground—the chief heat conveyed to the north being by atmospheric currents. During the third stage, when the land began to rise, the elevation, they argue, was only partial, and accompanied by depressions, which permitted warm oceanic currents to penetrate further north, and thus gradually to dissolve the ice-sheet to its present limits. Others also, who seek for the solution in the peculiar distribution of sea and land, contend that the ice-mantle was never general over the northern hemisphere, but that it passed from area to area according to the set and direction of the polar sea-currents. Were the existing arctic current, say they, to pass down by the coasts of Norway instead of by the coasts of Labrador, the glaciers of Scandinavia would envelop the whole peninsula, and come down, as they do in Spitzbergen, to the very sea-shore. It is only necessary, these theorists contend, to arrange the polar currents and the warm currents (like the Gulf Stream), or, in other words, the northward and southward currents of the ocean, in such a way as to produce over certain areas glaciers and icebergs, and all the accompanying phenomena of the glacial epoch.

Those, on the other hand, who appeal to extraneous causes, maintain that the ice-sheet was general and contemporaneous, and could only be produced by forces affecting alike the whole of the northern hemisphere. In their opinion, there was only a gradual advance, and as gradual a departure, of the ice-epoch. And this advance and departure may have been brought about by our planetary system passing through some colder region of space; by some secular or recurrent diminution of the sun's heat; by greater eccentricity of the earth's orbit; or by some secular deviation in the earth's inclination, depending upon change of centre of gravity, or the precession of the equinoxes. In other words, the causes they seek to establish are planetary and not terraqueous, and such as have already occurred, and will yet occur, in the history of our globe. Of course, before such hypotheses can be entertained, astronomers must admit their possibility, and find in the existing ordainings of nature some proofs that such great periodic changes are still in process of fulfilment. It is true that we are not without evidence of colder and warmer periods over the same latitudes in the geological history of our planet. It is many years ago[*] since we first drew attention to the occurrence of colder and warmer cycles—the Cambrian, Old Red, Permian, Cretaceous, and Boulder-Drift, being the *cold*, and the Silurian, Carboniferous, Oolitic, Tertiary, and Recent, being the *warm;* but we still need further corroboration of this, and it were the wisdom of geology to exhaust the operative causes that lie within our knowledge in the earth before seeking for extraneous causes that are placed in the mean time beyond our probation. In either case we cannot conceive of colder and warmer cycles occurring at random; and whether they depend upon the earth's own ordainings, or upon her wider planetary relations, they must be obedient to an orderly law of time, and will occur in the future as they have already taken place in the past.

Of course, before the origin and history of the glacial epoch can be fully expounded, much exact and prolonged labour must be devoted to the subject. The sequence and superposition of the clays, gravels, boulders, and silts, require more thorough examination over wider areas than those of the British Islands. The southerly limits of these accumulations in Europe, Asia, and America require to be determined with greater precision; as well as the question of their altitude, and their dispersion over eastern and central Asia. We have also to learn how far the southern hemisphere (in Patagonia, Australia, and New Zealand) has

[*] See 'The Past and Present Life of the Globe,' p. 189-191.

ever undergone a similar glaciation; and if so, whether it took place, as in the northern hemisphere, immediately preceding the current era.* These and similar questions must receive satisfactory answers before a generally approved theory can be hoped for; and geology, in this as in other instances, will best attain her end by a diligent accumulation of facts, and the widening of her field of observation.

Such is a hasty, but, we trust, not unintelligible Sketch of the glacial or ice epoch—that strange period in recent geology when the frost-giant that now reigns supreme within the polar circles laid his iron grasp on the northern hemisphere down to the 40th or 42d parallel of latitude. How or by what means this crisis was brought about—whether by some peculiar distribution of sea and land, or by some great secular recurrence in the earth's planetary relations —theorists are not agreed; but there is no difference of opinion as to its existence, and none as to the long continuance of its sway. Its presence is visible on every hillside and in every glen in the British Islands, Northern Europe, and North America; in the rounded eminences, in the polished and striated and grooved rock-surfaces, in the moraine-like mounds of gravel that bar the glens, in the huge rounded and striated blocks that lie scattered over the ground, and in the thick tenacious bouldery clay that envelops so much of the lower and level tracts of the country. No known agent, save ice, could have produced these appearances—ice on land, and ice on water; ice, in fact, such

---

* From recent reports by the provincial surveyors of New Zealand, as well as from Mr Darwin's well-known descriptions of South America, it would appear that the southern hemisphere has been subjected to a similar phase of ice-action. The further investigation of this as to contemporaneity with that of the northern hemisphere, would materially assist in the framing of an acceptable hypothesis.

as we now behold on the higher mountains of the globe, and within the arctic and antarctic regions.

We have arranged the period into three stages—the *first*, when the pre-glacial land (somewhat higher than the existing continents) began to receive the ice-sheet; the *second*, during which the ice-bound land subsided to the extent of 1800 or 2000 feet; and the *third*, during which the land was step by step re-elevated, and the ice gradually disappeared.* Each of these three stages must have left its own proper impress; but the second has obliterated so much of the first, and the third so much of both, to say nothing of what has been subsequently effaced by frosts and rains and rivers, that it is always extremely difficult, and often impossible, to discriminate their results. Hence the great difficulty of reading aright the phenomena of the glacial epoch; and hence the conflicting views entertained by geologists respecting their origin and arrangement. This much, however, is certain, that the pre-glacial or pliocene land-surfaces, wherever they are found, contain fossils; that the first stage of the ice-epoch is characterised by boulders little removed from their parent rocks, by finely glacialised rock-surfaces, and by the true boulder-clay or " till " of Scottish geologists, and is always unfossiliferous; that the second stage is characterised by re-assorted clays, by more rounded and widely-dispersed boulders, and is also

---

* While, for the sake of distinctness, we thus divide the ice-epoch into three great stages, it must be borne in mind that there may have been minor and local oscillations of sea and land during each successive stage. Since the close of the glacial period such oscillations have taken place more than once in our own islands, as proved by the " submarine forests " that occur at so many places along the existing coasts—these forests, now partially under the sea-level, having evidently grown at a higher elevation, been submerged to receive the silts that now cover them, and again upraised to their present levels. Such minor oscillations tend to complicate, but they do not obliterate, the broader phenomena of a period.

unfossiliferous; while the third stage has more moraines,
ridges of sand and gravel, terraces with occasional shells,
and finally, in the lower levels, the silty clay or "brick-clay,"
containing boreal shells, star-fishes, bones of seal, whale,
northern ducks, and other kindred remains. The local
differences may not be always ascertainable; the general
order above sketched is unmistakable throughout the
British Islands.

Cold and dreary, and inimical to life, as the ice-epoch
must have been, it has left its impress on every foot of the
surface to which its limits extended. The rounded outlines
of our hills, the gentler mouldings of our glens, the scoop-
ing-out of many of our higher lake-basins, the undulating
gravelly surfaces of our broader valleys, the terraciform
southern and south-eastern slopes of our mountains—nine-
tenths, in fact, of that which gives character and colour to
our northern scenery—are the direct results of its long-continued sway. Much has no doubt been since obliterated
by the frosts, rains, and running waters of the current era,
but the broader features chiselled out by the ice-epoch still
remain, reminding the spectator at every turn of its presence, and the long continuance of its power.

RECENT FORMATIONS.

As the glacial epoch, with its bouldery clays and gravels,
formed a limit to the tertiary system—over a large portion
of the northern hemisphere at least—so within the same
latitudes it constitutes an equally decided basement for
what are usually termed the Post-tertiary, or recent formations. Of course, like other appellations in geology having
reference to time, these terms are merely relative, embracing
accumulations that have taken place within the current
century, and others that may have been formed fifty or a
hundred thousand years ago, but still *recent* when compared
with those of the tertiary and other preceding systems.
This is all that is meant by the title at the head of the
present Sketch ; and whether we adopt Quaternary, Post-
tertiary, or Recent—designations all employed by geologists
in describing these formations—it matters little so long as it
is understood that the events referred to have taken place
subsequent to the glacial era.   These events, recent though

they be, present a curious but difficult chapter in world-history; curious as displaying more clearly than the older formations their whole origin and progress, but, like modern human history, difficult of narration, from the exuberance and nearness of the details. Approaching our own times, their interest is proportionally increased, and he who understands them aright cannot fail to catch by reflection a clearer insight into the cycles and systems that went before. In their origin and formation we see a repetition of the origin and formation of all the older formations, hence their instructiveness as a study ; while in their superficial dispersion they become the immediate source of sustentation to the plants and animals that inhabit the terrestrial surface.

Arising from the operations of waste and reconstruction described in Sketch No. 2, these Recent, or, as they are sometimes termed, Superficial Accumulations, will be as multifarious as the agencies concerned in their formation ; and hence perhaps the most intelligible way of treating them is to arrange them according to the agents more immediately concerned in their production. In this way we will have *Fluviatile* formations, or those arising from the action of rivers; *Lacustrine,* or those formed in lakes ; *Estuarine,* in estuaries; *Marine,* in seas; *Chemical,* arising from chemical action ; *Organic,* from the growth and decay of plants and animals; and *Volcanic,* from the internal fire-forces of the globe. There will be older and younger, of course, among these different formations—some so old as to imbed the remains of plants and animals no longer inhabiting the same localities, and others so recent as to belong entirely to the current age, and indeed to be still in process of formation. To display them, whatever their age, in intelligible order, is the object of the present Sketch ; and he who bears in mind the operations of waste and reconstruction described in a previous paper, can have no difficulty in following the narration of this, the most recent chapter in geological history.

Among the most obvious of Recent Formations are those produced by the action of rain and rivers. Whatever the winds and rains and frosts loosen and disintegrate, the stream carries onward and downward to lower levels. Were there no great rivers, the debris worn from the mountains would accumulate mainly along their bases, but the runnels gave it to the streams, the streams by their union to the river, and the river carries it forward to be scattered over the plains, to be deposited in lakes, or borne out to the depths of the ocean. Geologically speaking, what is strictly *Fluviatile* is laid down by the streams and rivers along their courses; and there is not a river in the world that does not present, in some portion or other of its course, patches of meadow-land, holmes, dales, and other flats, that have been formed by the debris carried down by its current. These alluvial flats are generally very heterogeneous in their composition—loamy and clayey silts, sand, gravel, and shingle, with here and there the imbedded but often imperfectly preserved remains of terrestrial plants and animals. In course of ages, as the river deepens its channel, and cuts its way from side to side down the valley, the older of these flats will stand higher and higher above the current; and thus it is that along most rivers there are sets of *terraces* marking the heights at which they formerly ran, and the levels over which they spread their inundating waters. It is usual to arrange these terraces into *higher river-gravels* and *lower river-gravels*—the former of vast antiquity, and rarely containing organic remains, and the lower of more recent origin, and containing the remains of plants and animals, some of which have long since become extinct in the regions where their relics now occur. It is from the

lower and middle of these terraces in Britain, France, and
other European countries, that the bones of the mammoth
and woolly rhinoceros have been exhumed, along with the
flint implements of rude and primitive races ; and were the
river-deposits of the other continents geologically examined,
there is no doubt they would exhibit in a similar way a gradation from the events of the current century back to those
bordering on the tertiary epoch. As it is, these river-deposits
play an important part in the physical geography of every
country, their rich and well-watered surfaces presenting the
finest fields, whether for forest-growth, pasture, or cultivation.
We have only to name the principal rivers of the world to
recall to the geographical reader the alluvial expanses that
mark the most of their courses, and these in magnitude
according to the volumes of the respective rivers and the
flatness of the country through which they flow.   And
even where magnitude is not concerned, these river-deposits
are not without their importance.   Every gully from the
mountain-side has its. terminal spit of sand and gravel,
and it is often in such deposits, worn from the cliffs and
veins above, that we find the most abundant and readiest
supplies of the gems and precious metals—as witness the
diamond-conglomerates of Brazil and India, the tin-gravels
of Cornwall, and the gold-sands and shingle of California,
Columbia, Australia, New Zealand, and the Oural.   These
auriferous sands and gravels are but the debris of the older
mountain-cliffs disintegrated by the frosts and rains, and
carried down by the streams to the lower "creeks" and
"gullies," where, accumulating for ages, they are often of
great thickness, and carry us, in many instances, far back
even into the tertiary period.   Many of the Australian
gold-gravels, indeed, are surmounted by thick overflows of
basaltic lava, and as volcanic agency has long ago ceased to
manifest itself in these regions, these show at once the vast

changes that must have taken place, and the ages that must have elapsed since the gravels and sands were piecemeal worn and washed down from their parent formations.*

Closely associated with these river-deposits, and sometimes indeed undistinguishable from them, are the *Lacustrine* or lake deposits, that occupy so many of the alluvial expanses in our plains and valleys. The tendency of every lake fed by running streams is to become shallower and shallower from the sediment deposited in its basin by these inflowing waters. Every stream protrudes its little delta of silt and sand, fresh-water shell-fish accumulate layers of marl, and aquatic plants contribute their annual quota of growth and decay. By-and-by the shallow lake becomes a stagnant morass, and in process of time, partly by surface plant-growth, and partly by the deepening of the outflowing stream, the morass is converted into meadowland. A large portion of all the " straths " of Scotland and the " dales " of England are of lacustrine formation; and we have only to watch the cutting of any main drain

---

* The following is a section of the Uralla gold-field, as given to us some years ago by Mr W. Cleghorn of that district :—

Red rich soil,	5 feet.
Stiff red clay,	5 ,,
Mottled clay—volcanic ashes,	20 ,,
Basaltic lava,	35 ,,
Brown laminated clay,	5 ,,
Loose sand (decomposed quartz and granite),	2½ ,,
Black peaty clay, with numerous leaves and stems,	6½ ,,
Loose sand (decomposed quartz and granite),	2 ,,
Finely laminated reddish clay,	1 ,,
Loose sand (decomposed quartz and granite), with numerous crystal-pebbles and *a little gold*,	15½ ,,
Fine reddish clay,	½ ,,
Loose sand (decomposed quartz and granite), with numerous pebbles—*the main gold deposit*,	4 ,,
	102 ,,

Granite, water-worn surface, with large granitic boulders.

through their subsoils to be convinced of the truth of this origin. A thick layer of vegetable or peaty soil, followed by beds of silty sand, marl, and clay, imbedding the bones of deer, oxen, and other animals, with the remains of an occasional tree-canoe, clearly bespeak their lacustrine formation, and point to the time when the wild animals of the country were mired in their muds, and the primitive inhabitants paddled across their waters. Now how changed! the site of the former lake green with the richest pastures, or waving with luxuriant corn-fields! As with the straths and dales of Britain, so with a large proportion of all the inland plains and valleys of the world. Many of them are but chains of silted-up lakes converted into dry land, partly by the process of silting or filling up, and partly by the main stream of the valley cutting deeper and deeper its channel, and thus affording a more thorough drainage for the whole. Lacustrine formations, though occurring in the same plain with those of fluviatile origin, are in general readily distinguishable by their finer sediments, greater regularity of deposition, the occurrence of beds of shell-marl and peat, and the more perfect preservation of their organic remains. These remains are often of great antiquity, ranging from the time of the mammoth, great Irish deer, and species of oxen that have been long extinct, down to the pile-dwellings of our Celtic or pre-Celtic ancestors, who betook themselves for safety to their waters, and erected artificial mounds for their habitations, where nature had not provided the necessary "inches" or islands.* Even since the time of the

---

* Since 1854 these lake-dwellings or pile-dwellings (known as *pfahl-bauten* in Switzerland, and *crannoges* in Ireland), have received much attention from archæologists and geologists. These dwellings occur in existing lakes, as well as in bogs and marshes which were formerly the sites of lakes, and seem to have been erected on piles driven through the water, or on mounds partly formed of stones, wood, and other debris. They have been found in Switzerland, Ireland, Scotland, and

earlier Celts hundreds of shallow lakes and morasses have been converted into dry land; and the process still goes forward, accelerated in all civilised countries by the incessant operations of man.

Still associated with river-action, but necessarily separated from strictly fluviatile deposits, are those *Estuarine* formations which occupy extensive areas in almost every region of the globe. Wherever a river discharges itself into the sea by a broad mouth, or by many mouths, and in particular where the tidal influence pounds back the river-water and runs for some distance inland, sandbanks and mud-shoals have a tendency to accumulate. In process of time the banks and shoals become islands, and by further accretion and union the islands are converted into *deltas.* In this way most of our larger rivers present deltic flats or estuarine formations, and these must have been slowly accumulating since sea and land received their present relative configurations. These accumulations, consisting mainly of debris borne down by the river, but partly also of tidal sediments, will imbed marine as well as terrestrial and

other European countries, and point to a time when the early inhabitants betook themselves to this style of habitation for purposes of defence and protection. In some instances, as in the Swiss lakes, the piled area is of considerable extent (forming an aquatic village, as it were), and connected with the shore by a piled way or causeway. In the older pfahl-bauten the implements are chiefly of *stone,* and associated with the cast-away bones of the deer, boar, and wild-ox; in those of intermediate age, *bronze* implements prevail, associated with the bones of the domestic ox, pig, and goat; while in the more recent, *iron* swords and spears have been found, accompanied by carbonised grains of wheat and barley, and with fragments of rude textures woven of flax, bast, and straw. The more recent seem to have been immediately anterior to the great Roman invasion of northern Europe; the more ancient may be several thousand years older than that event. Those curious in these matters may consult Dr Keller's 'Lake-Dwellings of Switzerland,' as translated and arranged by Mr Lee, 1866.

fresh-water organisms, and thus they are regarded as estuarine or *fluvio-marine*, in contradistinction to those of strictly *fluviatile* or *lacustrine* origin. The low-lying deltas of the Mississippi, the plain of Lower Egypt, the jungle-swamps of the Niger, the sunderbunds or mud-islands of the Ganges and Irawaddy, and the alluvial plain of China, are familiar examples on a great scale of these estuarine or deltic deposits; but as with these, so with almost every other river that discharges its waters into the ocean. The magnitude of estuarine formations is one of the most notable features in the geology of the current epoch, and this magnitude is increased by a twofold process which the reader would do well to consider.

In the first place, the delta that makes its appearance as dry land may form but a small portion of the sedimentary matter borne down by a river, the greater portion being carried forward and projected, as it were, over the bed of the ocean. An estuarine formation is thus partly *sub-aërial* and partly *submarine*, and it necessarily requires a long and gradual process of silting to convert the submarine into sub-aërial. But during the oscillations or crust-motions to which the earth is subjected, it frequently happens that a whole island or portion of a continent is gradually upraised, and thus the submarine portion of an estuary may be upheaved into dry land, and this altogether independent of the slow and ordinary process of silting. An uprise of fifty feet would convert a large portion of the Yellow Sea into a lower terrace of the great Chinese plain; and by a similar uprise thousands of square miles of Mississippi swamp would assume the character of fertile prairie-ground. To such upheavals much of the "carses" of Scotland and "levels" of England are no doubt due; for though wholly composed of river and marine silts, their final conversion into dry alluvial plains has been more a matter of terrestrial uprise than of sedimentary accumulation. As with the estuarine plains of our own islands, so to a great extent with those of other regions : they owe their accumulation wholly to silt and sediment, but their conversion into dry land, partly to silting, and partly to terrestrial upheavals.

Whatever their mode of accretion, the composition of these estuarine formations is much the same in every region : mud-silts, clays, sands, gravels, drift-wood, shell-beds, peat and swamp earths—the whole being usually surmounted by loamy, vegetable soils of extraordinary fertility. Their imbedded remains are partly terrestrial, partly fresh-water, and partly marine ; and these, of course, will differ according to the latitudes in which the estuary occurs, and the regions through which its affluent rivers flow. Thus the Mississippi will sweep down the terrestrial and fresh-water spoils of temperate North America, the Amazon those of tropical South America, the Niger those of Equatorial Africa, and the Ganges and Irawaddy those of subtropical Asia. Every estuary, in fact, is characterised by its own fossil flora and fauna, and these of varying antiquity, from the spoils swept down by the latest land-flood or deposited by yesterday's tide, back to the confines of the glacial epoch, if in the higher latitudes, and it may be to the tertiary itself, if occurring in intertropical regions. In the estuarine silts of our own islands, for example, we pass through every gradation of antiquity, from the plants and animals now flourishing around us, back through those which, like the bear, wild-boar, wolf, and beaver, have long since been extirpated, and from these backwards still to the seals, whales, and boreal shells that inhabited our firths and estuaries in times immediately post-glacial. The reader may readily trace this gradation in the estuarine deposits of the Clyde, Forth, Tay, or any other of our larger rivers. We take, for example, Stratheden in Fife (locally the "Howe" or hollow of Fife), and there, within a distance of twenty miles, we have first the sand-banks and sand-drift now in course of formation along the outer estuary; a few miles inland a greyish "carse-clay," with antlers of deer and bones of oxen, overlying a peaty layer of forest-growth replete with the stumps and trunks of the oak, pine, hazel, alder, birch, and willow, marking an oscillation of the land;* and still further up the strath, extensive deposits of sand and brick-clay, containing the remains of whales, seals, northern sea-birds, arctic shells, and star-fishes, usually regarded as immediately post-glacial.

Of Recent Formations, the next that fall to be considered are the *Marine*, or those accumulated in seas, and whose imbedded remains are chiefly of oceanic growth and habitat. Of course much of the sediment deposited in the ocean is brought down by rivers from the surface of the land, but we refer here to the areas in which it is collected, and the manner in which it occurs. Marine silt (mud, sand, and miscellaneous debris) is formed in every sheltered bay and recess of the ocean, where, under the influence of winds, waves, and tides, it gradually accumulates till it banks out the water, and is converted into tracts of low-lying alluvial land. The *fens* of Lincoln, and the *polders* of Holland, are familiar examples in our own seas; and as these have been formed, so have similar flats along every sea-shore favoured with the necessary shelter and the necessary tidal sets. In many areas it will

---

* This submarine forest, which is well exposed at the railway bridge across the affluent stream of the Motrie, is at the same level and of a similar character with those occurring on the Tay, Forth, Humber, the coast of Lincolnshire, Devonshire, Lancashire, and other localities.

be difficult to distinguish between estuarine silts and silts
that are truly marine, but in all bays and recesses where
there is no great entering river, the formations may be
regarded as belonging exclusively to the sea.   Besides
these *Littoral* or shore-formed silts, there are submarine
shoals and banks accumulating far from land, and sediments collecting in the stiller depths of the ocean.   Every
chart bears witness to the numbers of these shoals and
banks, and in certain areas every throw of the sounding-
lead brings up evidence of the vastness of these *Pelagic* or
deep-sea deposits.   On wave-washed shoals the accumulations are usually sand and shingle; but along the deeper
sea-bed they consist of slimy muds—the " oaze " or " ooze "
of the navigator.*   And were such pelagic deposits upraised into dry land, they would rival the older formations
alike in their geographical extent and in the diversity of
their mineral composition and organic remains.

Besides these truly marine sediments, there is another
set of accumulations of vast extent and peculiar interest,
which are partly marine and partly Æolian or wind-formed.
We allude to the sand-dunes, sand-drift, or links, which
border most of the bays in our own islands, and indeed the
bays and sea-shallows of every other country.   Originally
formed in the sea, the sands dried during neap-tides are

* Speaking of this mud, Captain Dayman, in his ' Deep-Sea Soundings,' says : " Between the 15th and 45th degree of west longitude lies
the deepest part of the ocean between Ireland and Newfoundland, varying from about 1500 to 2000 fathoms, the bottom of which is almost
wholly composed of the same kind of soft mealy substance, which, for
want of a better name, I have called *oaze.* This substance is remarkably
sticky, having been found to adhere to the sounding-rod and line through
its passage from the bottom to the surface, in some instances from a
depth of more than 2000 fathoms." On microscopic examination this
oaze was found to consist for the most part of foraminiferal organisms,
there being about 90 per cent of calcareous, and only 10 per cent of
siliceous matter; the mass when dried greatly resembling chalk in colour
and consistency.

drifted inland, and piled up in fringes of dunes or hillocks beyond the reach of the waves. By this process, continued year after year, and century after century, the sea is gradually banked back, and large sandy tracts created, terrestrial and wind-blown above, but marine and water-drifted below. We once watched the sinking of a well in the sandy tract that stretches between St Andrews and the Tay (Pilmoor and Tentsmoor Links) : the first fifteen or twenty feet were through wind-blown sands and thin layers of vegetable soil ; the remainder, to a great depth, was through masses of sand, shells, gravel, and shingle, the drift of St Andrews Bay when its waves rolled several miles farther inland than they do now. As with this instance so with others, whether along the shores of Holland, the Landes of Biscay, or the Bights of Western Australia. These sand-dunes, from narrow fringes of a few acres to rolling expanses of many square miles in extent, may be witnessed along the seaboard of every region, one of the most striking examples in Europe being the " Landes de Bourdeaux " (stretching southward from the mouth of the Garonne along the Bay of Biscay, and onwards towards Bourdeaux), with their shifting sandhills along shore, their dense forests of sea-pine farther inland, and beyond these artificially planted shelters vast expanses of heathy undulating sheep-runs.

In addition to these littoral silts and pelagic sediments, there is another class of marine deposits which have of late years much engaged the attention of geologists, partly from their varying and vast antiquity, and partly from the evidence they afford of repeated elevations of the land. We refer to those " raised beaches " or " ancient sea-margins," whose terrace-like flats are found fringing the seaboard of almost every region. Were our own islands to be upraised to the extent of twenty or thirty feet, the present shores (unless where steep and rocky) would form a low fringing terrace, here composed of gravel and shingle, there of sand and shelly debris, and in another portion of silty sediments. So it is with these ancient beaches; their composition is quite analogous, and their organic remains differ with their relative antiquities, or, which is the same thing, with their elevation above the existing sea-level. Such ancient sea-margins, at various heights (9, 25, 40, 63, and 120 feet), are very obvious along our own coasts, and, as already adverted to under "Crust-Motions" (Sketch No. 3), at still greater elevations (400 or 600 feet) along the shores of Scandinavia, Spitzbergen, Siberia, Greenland, and arctic North America. In South America also, both on the Atlantic and Pacific sides, similar terraces are frequent and boldly marked—those in Peru and Chili containing the "salinas," so valuable as the repositories of the salts of soda, potash, and other kindred substances.* Wherever they occur these ancient beaches notch the earth's surface like a great scale of time — descending from the glacial epoch down to the present day; all that is wanting being the numerical expression of their successive stages in years and centuries.

* The most important of these saline deposits (scientifically as well as commercially speaking) are those of Iquique in Peru. From six to fourteen leagues from the coast, and running parallel with it through the province, at an elevation of 3000 feet or thereabouts, is the pampa of Taramugal. This plain or pampa has evidently been a sea-lake, and in all likelihood the result of elevation by volcanic agency. There are other minor terraces or old sea-flats between the main pampa and the sea, but that of Taramugal is the most important and productive. It consists in some parts of many feet in thickness of sand indurated with salt, soft sand with crystals of nitrate, and true caleches of concreted nitrate of soda and stony debris. The other salts found in the deposit are chloride of sodium (common salt), biborates of lime and soda, sulphates of lime and soda, magnesian alum, &c.. Iodine also exists with the nitrate, and throughout the *calacheros* traces of boracic acid have been found in the water—the whole pointing unmistakably to the marine origin of the deposits.

The next class of Recent Formations that falls to be
noticed, embraces all such as arise more immediately from
Chemical actions and reactions.   It is true that chemical
changes are incessantly taking place in every formation,
whether of aqueous, of organic, or of igneous origin ; but
we allude in the present instance to those which, like tra-
vertine, sinter, bitumen, and the like, arise from deposition
and exudation by mineral springs and other causes chiefly
chemical.   Every one must have observed how the " petrifying spring" of ordinary language incrusts the stems,
leaves, and stones that lie along its course, the limy incrus-
tation thickening with time, and varying in magnitude
according to the volume of the spring, and the amount of
lime held in solution in its waters.   To this set of deposits
belong the *travertines, calc-tuffs,* and *calc-sinters* of the
mineralogist ; and the *stalagmites* and *stalactites* found incrusting the floors and depending from the roofs of caverns.*
These calcareous masses are found of all ages, from the
incrustations of the present century to the old cave-floors
imbedding the stone implements of primitive men and the
bones of extinct mammalia—cave-lions, cave-bears, hyænas,
mammoths, Irish elks, reindeer, and others, dating back
even to the tertiary epoch.   As with calcareous depositions,
so also with those of a siliceous or flinty nature, which arise
generally from hot springs, such as those of Iceland, the
Azores, California, New Zealand, and other regions.   These
siliceous tufas and *sinters* occur on a less extensive scale,
but they accumulate quite in the same manner, and imbed

---

* *Travertines* are light concretionary limestones deposited from waters
holding lime in solution, like those of the Arno and Tiber, hence *Tibertinus,*
Travertinus ; Sinters (Ger. *sintern,* to drop) are compact calcareous and
siliceous incrustations ; *tuffs* or *tufas,* on the other hand, are light and
porous ; *stalactites* (Gr. *stallaso,* to drop) depend from the roofs of calcareous caverns ; while *stalagmites* (Gr. *stalagma,* a drop) are the more massive
incrustations that accumulate on their floors.

or silicify whatever organism comes in their way.   To this
class of deposits belong also all superficial accumulations of
saline substances (common salt, soda, nitrates of soda and
potash, borax, &c.), sublimations of sulphur, exudations of
bitumen and asphalt, and indeed all formations arising more
immediately from chemical action and reaction.   Compared
with *mechanical* accumulations, they are usually of limited
dimensions, but they are often of curious interest, and their
study throws a flood of light on many of the older phenomena of the rocky crust.

Among Recent Formations those of *Organic* origin—that
is, arising from the growth and decay of plants and animals
—are, perhaps, the most interesting and instructive.   There
is no phenomenon more familiar to British readers than
the *peat-mosses* which occupy considerable tracts in every
part of the country, and which, before the extension of
modern agriculture, made their appearance over still wider
areas.   These peat-mosses are entirely of vegetable growth
—mosses, aquatic plants, heath, and fallen forests ; and
imbed isolated trunks of oak, pine, and birch, as well as the
remains of man, oxen, deer, and other animals that have
been swamped in their boggier portions.   These accumulations are of all ages, from the growth of the current year
back to the very close of the glacial epoch ; and their extent and thickness depend entirely upon the nature of the
area (its flatness, moisture, and shelter) in which they occur.
We have witnessed a depth of thirty feet reposing on blue
lake-silt which enclosed the remains of the extinct Irish
deer; while in another district similar silts have been found
beneath a peat-covering of less than a third of that thickness.
We have seen Roman remains of the time of Agricola found
at a depth of twelve feet ; while in other tracts we have examined the hewn stumps of trees, the plank-roads, the lost

armour and coins of the same invaders, at a depth of not more than five feet, and this all that had accumulated since the time they marched their legions across the Lowlands of Scotland. As with the peat-mosses of Britain, so with those of Holland and the north of Europe generally, those of Canada, and North America, and indeed of all temperate and coldly-temperate latitudes. They are of all ages, of varying dimensions and thickness, and of all degrees of purity, from masses wholly and exclusively vegetable, to others more or less intermingled with earthy debris, or with the siliceous and ferruginous accumulations of microscopic plant-growths and animalcules. In warmer latitudes, swamp-growths, jungle-growths, and other kindred vegetable accumulations make their appearance, the vegetable mass in these cases being more macerated and decayed than in the true peat-moss, but still wholly or almost wholly of organic origin. These peats and swamp-growths are in reality the coal-formers of the present day (see Sketch No. 9); nor let it be thought that in the aggregate they are either of inconsiderable thickness or of limited surface. In the peat-bogs of Europe, 20, 30, and 40 feet are no uncommon depths; the tundras of Northern Asia, though less pure in composition, are of much greater extent and thickness; the " Dismal Swamp" of the Southern States of America is 40 miles long by 25 miles broad; the peat-moss of Anticosti is stated by Sir William Logan to be 80 miles long, with an average width of two miles; and those in the lake regions of sub-arctic America are of still more extensive dimensions.

As with vegetable accumulations, so with those of animal origin — the coral-reefs, shell-beds, foraminiferal deposits, bone-shoals, and guano islands of the current epoch. The most obvious of these are perhaps the *coral-reefs*—the slow but incessant secretions of myriad polypes within all the warmer latitudes and shallower depths of the ocean. Within thirty degrees on either side of the equator, and at depths within twenty fathoms, these polypes in numerous genera are perpetually piling up their beautiful calcareous structures; here encircling lagoons, there fringing islands, and in another area extending in long ridge-like barriers. Many of these coral-reefs are of vast extent — the great " Barrier" of New Holland being upwards of 1000 miles in length, and from 20 to more than 100 feet in thickness; and in all, the mass is essentially composed of coral structure, but intermingled more or less with shells, crusts, coral-debris, and other extraneous substances. There is nothing more marvellous than this enormous secretion' of rock-matter by the tiniest of agencies; nothing more overwhelming to the conception than the number of individual organisms concerned in the work! And yet as it is now, so it has been in all time past; the same agencies have ever secreted the surplus lime from the ocean-waters, and built their reefs much in the same dimensions and much after the same style of construction. As with coral-reefs, so with *serpula-reefs* (annelids that secrete calcareous cases); *shell-beds*, whether drifted or buried *in situ; bone-shoals* drifted by currents, or frequented by fishes, seals, and other creatures that die and accumulate in myriads; and with *guano-deposits*, the droppings of sea-fowls that have accumulated under a rainless sky for ages. All are alike of animal origin, and all are alike returning to the solid crust that which other agencies had dissipated and dissolved. The importance of these organic deposits—*serpula-reefs* 12 or 15 feet thick, like those of Bermuda; *shell-beds* many leagues in extent, like those of the Indian Ocean; *bone-shoals* like that lying in the North Sea between the Faröe Islands and Iceland; and *guano-deposits* 40, 60, or 80 feet thick, like those off the coast of Peru—are not sufficiently recognised by geologists. As a curious illustration, we may notice the island of Sombrero, one of the West India group, which is almost entirely of organic origin. This islet, so called from its resemblance to a "sombrero," or low-crowned Spanish hat, and situated about 130 miles east of Porto Rico, is about two and a half miles long, one-half to three-fourths of a mile wide, and rises from 20 to 30 feet above the level of the ocean. It is a barren rock, and appears to be entirely composed of the rich phosphatic mineral known in commerce as *Sombrero guano*. This substance imbeds numerous bones of turtles and other marine animals; and from its composition, which resembles bones deprived of their cartilage, it has been supposed (with every degree of probability) that the island was once a shoal swarming with turtles and other vertebrate animals, whose accumulated remains of ages have been cemented together, and gradually elevated above the ocean-level to the present position of the island. The history of such an accumulation is interesting not only on its own account as a Recent Formation, but from the light which it throws on the origin of osseous breccias belonging to earlier epochs.

Besides these more obvious operations of plants and animals, there are others equally extensive though less apparent. The very lowest forms of life—organisms that stand, as it were, between the confines of the vegetable and animal kingdoms, and which require the aid of the microscope to reveal their forms—are equally busy, in equally inconceivable numbers, in accumulating their calcareous and siliceous remains. The calcareous oaze or mud of the deep seas, extending for thousands of miles, and of unknown thickness, is, as we have seen under the head of *Marine* formations, chiefly composed of the minute shells of foraminifera; and many siliceous muds in lakes, estuaries, and even in the deeper waters, are composed in like manner of the flinty shields of polycistinæ and diatomaceæ. The foraminifers and polycistines belong to the animal kingdom, the diatoms to the vegetable; but all require the higher powers of the microscope for their study, and when observers speak of miles of diatomaceous earths, and thousands of miles of foraminiferal muds, the mind is utterly impotent to grapple with the conception of the numbers of individual organisms that must have contributed their quota to the aggregate amount. But so it often happens in nature, that the most gigantic results are brought about by the minutest agents, and by the most imperceptible stages—the main conditions being incessant activity and unlimited duration.

The last of the Recent Formations that fall within the scope of our Sketch are the *Volcanic,* or those that have arisen from the substances discharged by modern volcanoes. These formations will consist mainly of *lava* in its different varieties, of *tufa* or compacted dust and ashes, of *scoriæ* or cindery and slaggy matter, of *pumice, obsidian, lapilli,* and *agglomerates* that have accumulated by the washing together of heterogeneous volcanic products. The products of every active volcano are more or less an epitome of those of all the rest; for though some, like those of the Andean cones, consist mainly of light cindery discharges, the great majority are admixtures, in varying sheets and masses, of all the substances above enumerated. Now a shower of dust and ashes blown aloft and scattered over one or other side of the mountain; now an overflow of lava slowly wending its way for miles down the rugged slopes; now explosive discharges of slag and cinders; and anon some gigantic lava-stream enveloping the whole in its rocky mass—again to be overlaid by repeated successions of similar materials. Etna, Vesuvius, and Hecla may be taken as familiar instances of volcanic accumulations, and as these have added hundreds of feet to their height, and cast their discharges for miles around during the current epoch, so have all other active volcanoes been similarly adding to their altitudes and lateral dimensions. One has only to cast his eye over a map of Volcanic Lines and Centres to see what a large area of the globe must have received accumulations of this kind within modern times—partly as mountain masses piled up on land, partly as islands upheaved from the sea, and partly as submarine sheets that have flowed as lava, or been cast abroad in showers of dust and ashes. All along the Andes, Central America, Mexico, the West Indies, and the north-western shores of the New World, volcanic accumulations have taken place, and are still taking place, on a grand scale. The same may be said of the Aleutian Islands, Kamtchatka, Japan, the Philippines, and the East Indian Archipelago; while over the bosom of the Pacific, along the Atlantic islands of Africa, in the Indian Ocean, in the Southern Ocean (as New Zealand), and in the Northern (as Iceland), similar phenomena have marked the course of the Tertiary period, and must, if we can judge from their present displays, have materially contributed to the exterior crust of our planet. Indeed, as there is no other Recent formation more obvious in its character and mode of accumulation, so there is none more rapid and gigantic in its results— a few months being often sufficient to pile up hills of slag and scoriæ, or spread abroad sheets of lava hundreds of feet in thickness, and many square miles in extent.* But it is

---

* Sir W. Hamilton reckoned the current which reached Catania in 1669 to be 14 miles long, and in some parts 6 wide; Recupero measured the length of another, upon the northern side of Etna, and found it 40 miles; Spallanzani mentions currents 15, 20, and 30 miles; the stream that flowed from the Skaptar Jokul in Iceland in 1783 was about 50 miles

not to the mere volcano alone that we must look for the full effects of vulcanic energy. The earthquake and crust-motions alluded to in a former Sketch (No. 3) are in like manner ever busy in moulding and modifying the earth's exterior—the former fracturing and fissuring the rocky crust, and giving tenfold significance to the discharges of the volcano, the latter silently elevating the sea-bed into dry land, or submerging the dry land beneath the waters.

Such is a rapid sketch of the Post-Tertiary or Quaternary formations, which, though recent and superficial to us, will become old and deep-seated to the observers of future ages. Scattered over the land, and spread out under the waters, they are not only everywhere present, but of all dates, from the accumulations of the current century back to those that mark the close of the glacial epoch. Indeed, the main difficulty connected with them is to assign their respective dates, and fix a scale of chronology that will be at all generally applicable. Superposition can only be applied in very limited cases; and mineral composition is of little avail, as the older are often as loose and unconsolidated as the younger. The only satisfactory test is fossils, striking back from the existing flora and fauna of any locality to those that have been removed from that locality, and from these back to such as have become totally extinct. In this way we may speak of *Upper*, or those containing existing plants and animals; of *Middle*, or those characterised by plants and animals locally extirpated; and *Lower*, or those marked by organisms now wholly extinct. As regards Man, they

long, by 12 or 15 in its greatest breadth, and from 20 to 600 feet in thickness, according to the nature of the ground; while Dr Coan estimates the discharge of Mauna Loa (one of the Sandwich Island volcanoes) in August 1855, at 70 miles long, with a varying width from 1 to 5 miles, and from ten to several hundred feet in thickness.

may be arranged into *Pre-human* and *Human*—the former containing no traces of man or of his works, the latter being here and there characterised by such remains. And even this Human period may be conveniently subdivided into *Pre-historic* and *Historic*—the former, like the flint implements, shell-mounds, and cave-dwellings, dating back far beyond the reach of history, the latter coming within periods to which we can assign something like a date in years and centuries. But whatever the arrangement, we clearly perceive that some are very ancient—so ancient as to merge into the close of the glacial period,—and others very recent—so recent as to be still in progress of formation. Nor let it be imagined that, because recent, these accumulations are limited and insignificant. All that is necessary to make them rival the older formations in extent and thickness is *time*, and this is an element as unlimited in the future as it has been prodigal in the past.

During the deposition of the more ancient, the mammoth, woolly-haired rhinoceros, hippopotamus, cave-lion, cave-bear, hyæna, and great Irish deer, were inhabitants of Western Europe; herds of mastodon roamed over the river and lake flats of North America; the plains of South America were densely peopled by the megatherium, megalonyx, glyptodon, mylodon, and macrauchene; and Australia, with more gigantic forms of kangaroo and other marsupials. At this period we have no traces of man save in Southern and Western Europe * (the only tracts yet sufficiently examined); and there savage races seem to have lived in caves and wigwams, fashioned stone and flint implements,

---

* We do not lose sight of the fact, that implements fashioned out of the native quartzite of India have been found in the alluvial laterite of Madras and North Arcot, but merely hold it subordinate till the discovery of associated organisms enable geologists to form a more definite idea of the relative antiquity of these lateritic deposits.

and subsisted chiefly by hunting and fishing. During the deposition of the less ancient or middle formations the earlier fauna had in a great measure disappeared, and species of ox, deer, horse, wild-hog, wolf, bear, and other existing genera had taken their places. To this period belong the extinct species of ox, horse, and deer, which have been found alike in Europe and America; and perhaps also the gigantic ostrich-like birds of Madagascar and New Zealand—the æpiornis, dinornis, palapteryx, and their congeners. The men of Western Europe still fashion their stone and bone tools and tree-canoes; but traces of underground stone-dwellings, and pile-dwellings in lakes, with doubtful indications of metal implements, bespeak an advance and mark the first stages from savagery to civilisation. During the deposition of the more recent, the flora and fauna of every region have remained much as we now behold them.* A few general extinctions, like the dodo, solitaire, great auk, and rhytina,† have been recorded; hundreds of local extirpations and removals (like the original flora and fauna of our

---

* According to Professor Heer, the native flora and fauna of Switzerland have remained much the same since the time of the earliest lake-dwellings, while the cultivated plants and domesticated animals have passed into totally different varieties. If this observation be correct, it tends to show that organic changes are slow or rapid in proportion to the physical changes to which life is subjected, and that where the physical surroundings undergo slow and gradual mutations (which is the common course of nature) plants and animals may exhibit little variation for ages. It is thus that the specific changes recorded by Palæontology afford the strongest evidence of the incalculable lapse and length of geological time.

† The circumstances connected with the extinction of the dodo, solitaire, and great auk, are well known. The rhytina, a phytophagous Sirenian, discovered by Steller on Behring Island in 1741, is also considered as completely extirpated—the last individual having been killed in 1768. Unlike the manatees or sea-cows, the rhytina was edentate, having special bony palatal apparatus for the crushing of its food. Its sub-fossil remains (from 8 to 24 feet in length) are now eagerly sought after for our public museums, and one or two specimens, we believe, were exhumed in 1864.

own islands) have taken place under the aggressions of man; and man himself has risen through the successive stages of *stone, bronze,* and *iron,* to what we now behold him. We pass from pre-historic to historic times, but still there is no cessation. The agencies of nature are as busy now in moulding and remodelling the rocky crust as ever they were. From their slow and gradual operation we may fail to appreciate the results; to the future, however, they will appear in all their vastness and universality.

MAN'S PLACE IN THE GEOLOGICAL RECORD.

UNLIKE the periods of human history, those of Geology
have no definite expression in years and centuries. We
speak of eras and epochs, of cycles and systems, but these
are merely relative terms. They have no definite duration ;
the one merely precedes the other, and the larger may in-
clude many recurrences of the lesser within its limits. In
speaking of geological time this is all that is signified ; in
fixing the dates of geological events this is all that can be
fairly asserted. The Primary merely precedes the Secondary,
the Secondary the Tertiary, and the Tertiary the events of
the Current epoch. We may subdivide these greater stages
into narrower limits, and talk of Laurentian, Cambrian,
Silurian, Devonian, Carboniferous, Permian, Triassic, Oolitic, Cretaceous, Tertiary, and Recent rock-systems, and this
is no doubt restricting events to more precise bounds, but
it gives no definite idea of duration, nor tells us how long
the Chalk preceded the Tertiary, or the Tertiary the occur-

S

rences of the existing epoch. We can judge from its thickness, and the nature of its rocks and fossils, that one system took much longer time to accumulate than another, but we cannot venture, by any known method of computation, to say how long in years. All that we have to do with is relative time; and even in dealing with the current epoch, should we assert that certain events took place more than six thousand or eight thousand years ago, we are simply asserting a provisional opinion, and not maintaining a belief like that founded upon the written record of human history.

The geological record is thus relative and not absolute; and when we arrange it, as in the subjoined tabulation, into Primary, Secondary, Tertiary, and Quaternary, we are merely asserting a certain order of succession, and this succession not always clearly defined over certain areas. Indeed, it is often impossible to define the boundaries of the minor stages, portions having been removed by denudation, others overlaid by more recent deposits, and some being partially submerged beneath the waters of the ocean. Again, though the thickness of one formation may seem to have required a longer time for its accumulation than another of smaller dimensions, yet in the one case the rate of deposition may have been much more rapid than in the other, and the thinner may, after all, have required the longer period. Still further, though organic remains are most important aids, yet they are often absent from certain beds, or if there, these beds are not sufficiently exposed to investigation, and our information becomes in this way fragmentary and defective. Neither in sequence of events, nor in expression of time, does Geology lay claim to exactitude. Its cultivators are successfully labouring to complete the one, and they are hopeful of arriving at more definite terms in the other; but this is all in the mean time, and the following arrangement expresses the amount of their information :—

CAINOZOIC (*Recent Life.*)	{	Quaternary or Recent formations. Tertiary.
MESOZOIC (*Middle Life.*)	{	Cretaceous or Chalk. Oolitic or Jurassic. Triassic or Upper New Red Sandstone.
PALÆOZOIC (*Ancient Life.*)	{	Permian or Lower New Red Sandstone. Carboniferous or Coal System. Old Red Sandstone and Devonian. Silurian.
EOZOIC (*Dawn Life.*)	{	Cambrian. Laurentian.

In this arrangement the terms Eozoic, Palæozoic, Mesozoic, and Cainozoic, indicate the chronological stages having reference to the ascent of life; and Laurentian, Cambrian, Silurian, &c., those having reference to the different formations whose depositions mark the successive physical operations of nature. By this arrangement the geologist simply asserts that the Laurentian preceded the Cambrian, and the Cambrian the Silurian, but no opinion is expressed as to the amount of time required for the deposition of the Laurentian, or whether the Cambrian occupied a longer time in formation than the overlying Silurian. We may feel convinced, from the total thickness of a system, the alternations of its strata, and the succession of its fossils, that it occupied a much longer time in formation than another system; but this is not expressed in the above arrangement, which merely affirms a sequence from older to younger, and from the earliest ascertainable operations to those still taking place around us.

Such is the chronology of Geology—a chronology to which investigators endeavour to conform the rock-formations of the globe; and although the Chalk of one country, for example, may not have been exactly contemporaneous with

the Chalk formation of another region, still we know that
it stands intermediate between the Oolite and Tertiary, and
can therefore assign to it a place relatively to these formations.   In some region yet unexplored a whole suite of
strata may be discovered older than our Carboniferous, and
yet younger than our Old Red, and in such a case geologists would give the new formation a name, and place it as
intermediate between these two systems.   It would disturb
no established order, but merely render more complete the
sequence, like the interpolation of a hitherto unknown reign
in the dynasties of human history.   The geological record
is thus a thing of mere sequence—an inconceivable amount
of unexpressed time, during which certain events follow each
other in definite order.   How many ages have elapsed since
the first deposition of the Laurentian strata we cannot tell ;
how many centuries were spent in the formation of the
Coal-measures of any locality, we can only, estimating from
existing operations, offer the widest conjecture.   But we
can affirm with certainty, and this is a great point gained,
that one rock-system is younger than another ; that these
rock-systems follow in the order above given ; that according to our present knowledge invertebrate life preceded the
vertebrate ; that fishes preceded reptiles, reptiles birds, and
birds mammalia.*   We can also affirm, what it is the object of the present Sketch to prove, that as there has been
an ascent in time from lower to higher forms of life, so
Man, being the highest known creature, comes latest on the
geological stage, and that evidences of his existence are to be
found only in the most recent and superficial formations.

It will be seen from the preceding statements that the
geological record is avowedly indefinite and defective—indefinite, as it deals only with relative time ; and defective,
as many strata cannot be assigned to their proper positions,

* See tabulation of ascending orders, p. 29.

partly from the obscurities of superposition, and partly from the absence of typical fossils to connect them. But, while admitting this defect in details, it must not be imagined there is any uncertainty as to the broader features of the record, or that any new discoveries have ever been at variance with the great order of sequence which modern geology has established. Man, so far as every known fact tends to indicate, belongs exclusively to the Recent or Post-tertiary period. No remains of his kind, no fragment of his works, no traces of his presence, have ever been detected in earlier formations. But though this is admitted on all hands, the question still remains, at what stage of the Post-tertiary are traces of his existence first detected? Till recently the general belief has been that man's first appearance on the globe dates back, at the very most, to little more than six or seven thousand years; and so incorporated had this belief become with others of a more sacred character, that few, even though doubting, had the boldness to express a contrary conviction. Like the age of our planet, which was also at one time restricted to a few thousand years, the antiquity of man has become a question of science and reason; and well-informed minds are now prepared to admit that as the earth has existed for untold ages, so man, its latest creation, may have inhabited its surface for hundreds of centuries. The evidence is purely geological, and as such ought to be treated like any other problem in science, without bar or hindrance from preconceived opinion; or, as it has been well said by Bishop Tait, in his address to the Philosophical Institution of Edinburgh, " The man of science ought to go on honestly, patiently, diffidently, observing and storing up his observations, and carrying his reasonings unflinchingly to their legitimate conclusions, convinced that it would be treason at once to the dignity of science and of religion, if he sought to help either by swerving ever so little from the straight rule of truth." In investigating the antiquity of man we are dealing with a question of natural history, and are bound by the same methods of research as if we were treating of the history of the mammoth or mastodon. Our business as geologists is to examine the rock-formations composing the earth's crust, to note their imbedded organisms, and to fix their relative antiquities. Wherever the remains of man or of his works occur, there, we presume, has been his presence ; and though we cannot assign any definite date to the time of such occurrence, we are at all events entitled, judging from all the correlative circumstances, to say that it took place more than six thousand, ten thousand, or twenty thousand years ago. In other words, we are bound to deal with Man's antiquity as with any other question in geology ; and though our dates be merely relative, we can affirm the order of sequence, and arrive at some notion of duration from the rate of existing operations.

Abiding by these methods, we find the remains of man and of his works gradually receding from the historical into the pre-historic ages. In Southern and Western Europe—the only regions that have been examined with anything like geological accuracy—these remains occur in peat-mosses, in lake-silts, river-drifts, and cave-earths, and from their associated organisms we judge of their relative antiquities. If they occur along with the remains of the existing horse, ox, sheep, pig, and the like, we know that they are comparatively recent, and in all probability belong to the historic era. If, on the other hand, they are found accompanied by remains of extinct species of horses and oxen, we know they are of greater antiquity ; and if such horses and oxen are not spoken of in history, or represented in human monuments, then we are entitled to regard them as pre-historic. Or again, if they are associated with remains of the great

Irish deer, the mammoth, mastodon, woolly-haired rhinoceros, and other animals long since extinct, we feel assured that vast changes in physical geography have taken place since their entombment, and are entitled to assign to them a still higher antiquity. In fact we know that all changes in physical conditions, and all removals and extinctions of life, take place by slow and silent stages, and that the greater the difference between the existing and the extinct, the longer must be the time that has elapsed since their extinction. By methods such as these we can establish a scale of old, older, oldest; and there need be no more uncertainty about the results obtained by such methods than there is about the results obtained by the historian in modern, medieval, and ancient history.

Another method by which we arrive at notions of relative antiquity is by the implements and works of art that occur in recent formations, or accompany the remains of man. We know the phases of modern, medieval, Roman, Greek, Egyptian, and Babylonian art, and can assign something like a historical date to such objects and the accumulations in which they occur. We know, too, that man employs tools of wood and stone long before he learns the uses of the metals; and that he reduces the softer metals, and works in copper and bronze, long before he has acquired the mastery over iron and steel. In this way we speak of the ages of *stone, bronze,* and *iron,* the one preceding the other, and forming, as it were, a rude scale of time for the antiquarian and geologist. But while one nation may be working in iron, another more belated may be working in bronze, and a third, still more remote and savage, may be adhering to implements of wood and stone. To be of any use, this scale of stone, bronze, and iron, must be applied to the same district; and when so applied, archæologists are now pretty well agreed that it marks with considerable certainty the various stages of relative antiquity. Of course, were implements of iron ever found along with remains of mammoth and mastodon, the scale would be utterly worthless; but when stone tools invariably accompany the older remains, and those of bronze and iron those of younger and younger date, then we feel assured from this concordance of the implement scale with that of the animal that we have hit upon a pretty exact method, so far as Europe at least is concerned;* and it is by both of those modes that man's place in the geological record has been mainly determined.

It will be seen that in speaking of implements of stone, bronze, and iron, the geologist is trenching on the field of archæology, and the archæologist on that of geology. Both must, in fact, lend their aid in solving the question of man's antiquity; and whether it be by sepulchral barrows, by shell-mounds—the old feasting-stations of our northern ancestors—by pile-dwellings in lakes, or by flint implements in river-drifts, much the same kind of reasoning must be employed by both. A lake-dwelling, with implements of stone and bronze, may carry us no further back than the time of the Romans; while a tree-canoe, hollowed out by fire, and found under twelve or fourteen feet of river-silt, may take us thousands of years before Rome had a foundation. The inhabitants of Northern Europe may have lived on shell-fish and been wrapt in skins when the Pharaohs were clothed in fine linen and purple; but when we find

---

* Some archæologists divide the Stone Period into the *palæolithic* and *neolithic* stages—the former the age of rude stone implements, and when man shared the possession of Europe with the mammoth, the cave-bear, the woolly-haired rhinoceros, and other extinct animals; and the latter the age of polished stone implements, and when man began to domesticate the dog, ox, horse, and other existing mammalia. In this way we have four stages of pre-historic time :—1, the Ancient Stone age ; 2, the Newer Stone age ; 3, the Bronze age ; and, 4, the Iron age. For much interesting and well-condensed information on this topic, see Lubbock's ' Pre-historic Times,' 1865.

stone implements associated with worked horns of the great Irish elk and reindeer, and with bones of the musk-ox, mammoth, and woolly-haired rhinoceros, and these in silts and drifts that indicate great physical changes in the geography of Europe, then we may rest assured that these monuments are *pre-historic* and of unknown antiquity. We have no indication in history that the mammoth, rhinoceros, or Irish deer were inhabitants of Southern and Western Europe; nothing either in history or tradition that points to the time when the reindeer and musk-ox roamed in the latitudes of France and England. It is true that natural events are rarely noticed in ancient history, and especially those of slow and gradual occurrence like the facts of geology; still it may be safely asserted that during the historic period none of the animals above referred to were inhabitants of the southern and western portions of our continent. Whatever the date of these stone implements, and their associated mammoth and rhinoceros remains, they clearly belong to pre-historic times; and the question is thus narrowed to the relative antiquities of certain events which occurred far beyond the reach of the oldest history and the remotest traditions.

In dealing with pre-historic monuments, we may adopt either the methods of the archæologist, who founds chiefly on the comparative rudeness and simplicity of the relics, or those of the geologist, who looks mainly to the superposition of the beds in which the relics occur, or those of the palæontologist, who argues from the specific differences of the flora and fauna; or we may adopt a mixed method, and reason from all that archæology, geology, and palæontology supply. Adopting this latter plan, we reason from the lake-silts, peat-mosses, and deltic deposits containing stone implements and tree-canoes, associated with the bones of extinct varieties of the horse and ox, back to similar depo-

sits and cave-earths imbedding ruder implements and remains of the Irish deer, reindeer, and musk-ox, and from
these again to deeper river-gravels and brick-earths containing implements still simpler in fashion, and associated
with the relics of mammoth and rhinoceros. Considerable
changes in the physical geography of Europe must have
taken place (as these silts and peat-growths imply) since
the time of the primitive horse and long-fronted ox; still
greater must have taken place since the reindeer and musk-
ox found a suitable climate in the latitude of France and
England; and greater still since the mammoth roamed in
the pine forests and over the plains of the same regions.
Admitting the changes, the question remains, How shall
we estimate the lapse of time required for their fulfilment?
If they are changes of a physical kind, we estimate according to the rate at which similar changes are taking place
at the present day; if of a vital kind, by the rate at which
extinctions and creations seem to have been effected in
former epochs; and if of a kind involving the progress of
our own race, we know that civilisation in the long-run is
only arrived at, even under the most favourable circumstances, by slow and gradual stages.

Guided by these methods, the pile-dwellings in lakes (the
pfahlbauten of Switzerland and the *crannoges* of Ireland
and Scotland*) carry us back to the earlier Celtic times,
and may range from two to four thousand years, but clearly
they are not of the vast antiquity some archæologists have
imagined, and though pre-historic in Europe, may have been
contemporary with historical events in Egypt and Western
Asia. Estimated by the implement-scale, they belong alike
to the ages of iron, bronze, and stone, and mark the long
occupancy of South-western Europe by the same partially
civilised but gradually improving race. As regards the

* For an account of these Lake-dwellings, see note, p. 254.

shell-mounds (the *Kjökken-mödding* * of Denmark) and cave-dwellings of Belgium and France, they seem to indicate the presence of a pre-Celtic people, simpler in their mode of life, less civilised, and only acquainted with the use of implemènts in stone, wood, and bone. Smaller in stature than the Celt, round-headed, hunters and fishers, these pre-Celtic races never seem to have cultivated the soil, or to have settled down in fixed situations. Western Europe appears to have been their home before the Celts left the mountains of the East; and five or six thousand years ago may mark the date of their occupancy of the regions where now are found their shell-mounds, cave-dwellings, and kindred reliquiæ. Still earlier than these pre-Celts, Southern Europe to the shores of the Mediterranean, and Western Europe to the limits of the British' Islands, seem to have been occupied by a ruder but perhaps kindred race—the fashioners of flint implements, and the contemporaries of the reindeer, the mammoth, and woolly rhinoceros. Reindeer, hairy elephants, and woolly-haired rhinoceroses, in the latitudes of France and England, bespeak a severer climate than at present prevails, and under this boreal climate these rude races seem to have earned a scanty subsistence, by hunting and fishing along shore, by lake,

---

* Literally "kitchen-middens;" the name given by the Danes to certain mounds which occur along their sea-coasts, and which consist chiefly of the castaway shells of the oyster, cockle, periwinkle, and other edible kinds of shell-fish. These mounds, which have also been found on the shores of Moray and the north of Scotland, are from 3 to 10 feet high, and from 100 to 1000 feet in their longest diameter. They greatly resemble heaps of shells formed by the Red Indians along the eastern shores of the United States, before these tribes were extirpated. The "kitchen-middens" of Europe are ascribed by archæologists to an early people unacquainted with the use of metal, as all the implements found in them are of stone, horn, bone, or wood, with fragments of rude pottery and traces of wood-fires. All the bones yet found are those of wild animals, with the exception perhaps of the dog, which seems to have been domesticated.

and by river-side. And it is generally in such situations that their flint implements are found associated with the bones and tusks and horns of these extinct mammalia. But these implements (like those of Abbeville, &c.) are often found at great depths, and at altitudes above the levels of existing rivers, that prove the occurrence of great physical changes in these regions; and this, taken in conjunction with the extinction of the mammalia and the evident amelioration in climate, bespeaks a vast antiquity compared with the shell-mounds and pile-dwellings of the preceding races. A vast antiquity! but whether ten, twelve, or twenty thousand years, we have in the mean time no mode of precisely determining.

Physical changes proceed at rates too uncertain to constitute a scale of chronology, and we know too little of the law of vital development to found upon the duration and extinction of species. But if we may judge from existing operations, and if we may estimate from the specific changes in life now going on around us (and this with all the interfering influences imposed by man), then the time must be vast indeed since these primitive races were the inhabitants of Southern and Western Europe. We do not contend, like some, for thousands of centuries ; but we argue for triple or quadruple the amount that has hitherto been assigned to human chronology. Let us look fairly at the facts: the river-drifts, cave-earths, and lake-silts are, no doubt, very ancient, but there is nothing connected therewith that may not (computing by existing operations) have been accomplished in ten or twelve thousand years. Again, the mammoth, woolly rhinoceros, cave-lion, cave-bear, and cave-hyæna, are but species of existing genera ; and so little do they vary in general character from those still living, that their appearance at the present day would excite no marvel. The whole aspects and surroundings of these extinct mammalia are in truth geologically recent; and when we further consider the fresh condition in which some of them occur in the ice-gravels of Siberia, we are compelled to withhold from them an unlimited antiquity. It is a sound maxim in palæontology, that the greater the divergence of any species from existing species, the greater its antiquity; and founding on this rule, the mammoth, mastodon, and their huge congeners, cannot lay claim to the vast antiquity which many geologists have been so anxious to assign to them. Still, with all these facts and allowances, it must ever be remembered that the occurrence of hairy elephants and woolly rhinoceroses in Western Europe bespeaks a much colder climate than the present; and as changes in climate can only arise from great physical changes, great alterations must have taken place in the external conditions of our continent. Such changes are ever slow and gradual, and thus we are compelled to admit a high antiquity to the fashioners of these flint implements and their contemporaries, the mammoth and mastodon.

Indeed, the existence of a boreal climate necessitating shaggy coverings for the elephant and rhinoceros, would seem to carry us back to times immediately post-glacial— that is, to the time when the last traces of the glacial epoch were gradually being effaced by the advent of a more genial and equable climate. Were this the case, the appearance of man in Europe would be coeval with the earlier Post-tertiaries, and his antiquity much higher than the majority of geologists are yet prepared to admit. But his occurrence in Europe does not settle the question of his first appearance on the globe. On the contrary, the human race, in one or other of its varieties, may have existed for ages in Asia or Africa before it found its way to Western Europe, and, indeed, all that we know of language and ethnology seems to point to this conclusion. Before we can arrive at the absolute antiquity of man, or of his real place in the Geological Record, we must know more of the Asiatic and African Post-tertiaries, and more of the correlation of these to the Post-tertiary accumulations of Europe. We must also learn to deal with man as with other fossil genera, and instead of seeking for mere variations in skull and facial angle, we must be prepared to admit variations that amount to true specific distinctions. All animals in the history of the past, if they have existed long enough, break into varieties and species; and it will be a proof of man's comparative recentness, if we can discover no wider difference than mere varieties; but, on the contrary, it will be evidence of his higher antiquity, if zoologists can show that any variation, past or existing, is so great as to entitle it to be ranked as a specific distinction. Man may be the sole species of a single genus, but in this particular zoologists have departed from the true Baconian method, and dealt with man as if he did not belong to the same category of vitality with which it is the duty of their science to deal; and not till they have learned to treat him from a natural-history point of view, can we hope to receive from them anything like truly philosophical opinion.

As the matter stands at present, we have evidence of man's occupancy in Europe during the formation of the earlier Post-tertiaries, and during the period when the reindeer, musk-ox, hairy elephant, and woolly rhinoceros roamed over its surface. The existence of these animals in Western Europe betokens a somewhat boreal climate, and in all likelihood man gradually took possession of the continent as the climate began to improve on the gradual recession of the glacial epoch. Arranging the Post-tertiary system, as has been proposed, into *Mammothian*, *Reindeer*, and *Bovine* stages, we find man occurring at least during a portion of the Mammothian stage, and thus bespeaking for him a vast and venerable antiquity—unexpressed in years, no doubt, but not on that account the less certain in its existence and duration. But while man's place in the geological record belongs to the earlier Post-tertiaries in Europe, older varieties of his race may have existed for untold ages in the regions of Asia and Africa, from which in all likelihood the European branches were descended.* On the advent of the glacial epoch over the latitudes of Europe, the pre-glacial animals seem to have receded to southern and more genial climates, and again on its departure they appear, in some of their species, to have returned to the old areas. It was during this post-glacial return that man seems to have made his first appearance in Europe—a fisher and hunter, forming rude stone implements, and, so far as geology has discovered, very low in the scale of civilisation. But while Mammothian man was struggling along the river-banks of Europe for a scanty subsistence, other families of his race were in all proba-

* "It is not under the hard conditions of the glacial epoch in Europe," says Dr Falconer, "that the earliest relics of the human race upon the globe are to be sought. Like the Esquimaux, Tchukche, and Samoyeds on the shores of the Icy Sea at the present day, man must have been then and there an emigrant placed under circumstances of rigorous and uncertain existence, unfavourable to the struggle of life and to the maintenance and spread of the species. It is rather in the great alluvial valleys of tropical or sub-tropical rivers, like the Ganges, the Irrawaddy, and the Nile, where we may expect to detect the vestiges of his earliest abode. It is there where the necessaries of life are produced by nature in the greatest variety and profusion, and obtained with the smallest effort—there where climate exacts the least protection against the vicissitudes of the weather—and there where the lower animals which approach man nearest now exist, and where fossil remains turn up in greatest variety and abundance. The earliest date to which man has as yet been traced back in Europe, is probably but as yesterday in comparison with the epoch at which he made his appearance in more favoured regions."—*On the asserted occurrence of human bones in the ancient fluviatile deposits of the Nile and Ganges—Quarterly Journal of Geology,* 1865.

bility—we may almost say were undoubtedly—enjoying a higher civilisation in the sub-tropical and higher tropical regions of Africa and Asia. Were these Asiatic races of the same variety of our species as the Abbeville flint-formers, or did they, though enjoying a higher degree of civilisation, belong to some older but inferior variety? Much, indeed, in the matter of man's antiquity will depend · upon how this question is answered by subsequent discovery. If they belong to the same race, and there be no indication of any inferior species of our kind, in accordance with the great law of animal development, then, geologically speaking, man is of comparatively recent origin, and the question is narrowed to one or other of his existing varieties. Our own opinion is that, granting a law of development, the higher animals pass through fewer intermediate stages than the lower, and that, in man's case, species more closely related to the Quadrumana are scarcely to be expected. But while this may be true, it is equally certain that if there be any truth in geological development at all, the higher varieties must be more recent than the lower; and thus the white variety of man more recent than either the Red Indian, the Negro, the Malay, or the Mongol. And it is equally certain, according to any law of development, that the older and lower varieties must first pass away—a fact in wonderful accordance with the gradual disappearance of the coloured varieties before the spread of the white variety of our kind. Here, then, we have a twofold argument that may avail us in our researches—viz., the earlier appearance, and, conversely, the earlier disappearance, of the lower varieties of a species; and applying this to man, the coloured varieties, which are evidently inferior (whatever may be said to the contrary), must have long preceded the white, just as now they are passing away before it.

In this way we carry the antiquity of man—high as it may be in Europe—to a still higher antiquity in the other continents of the Old World, and which must be geologically investigated before any definite conclusion can be arrived at either as regards time or developmental descent.* The European men of the Bovine and Reindeer periods evidently belonged to the white or Caucasian variety, but we have no certain evidence whether the Abbeville flint-fashioners were of Caucasian, Mongolian, or other variety. To whatever variety they belonged, they were clearly of a date immediately post-glacial; though, could it be shown by craniology that they were of other type than the Caucasian, it would in our opinion be further proof of their high antiquity. If we are to pursue the subject of man's antiquity in Africa or Asia, this question of type must constitute one of the main elements of determination, for it would be outraging every principle in science to apply the test of variation and development to the other orders of life, and shrink from applying it in the solitary instance of man. Where we can prove by archæological means a high antiquity for man, let us adopt them ; where we can show the same result by geological methods, let us not neglect them ; but at the same time let us also value those palæontological doctrines of progression and development which have thrown so much light on the order and connection of vitality in general. If there be such a law of progression, man must be as amenable to it as the rest of creation, and whatever variation occurs in his race must be taken, along with other elements, as a measure of time and duration. We are aware that many geologists shrink from

* Since the above was written, we observe that implements of quartzite have been discovered in the lateritic formation of Madras by Messrs Foote and King of the Indian Geological Survey, thus opening the way to this new and much desiderated line of evidence.

this test of variation, and feel an uneasy tenderness when-ever the question of man's descent becomes involved in their researches and speculations. Truth, however, will never be attained by such weakness. In science, as in morals, error becomes only more deeply rooted, and bigotry more emboldened, the longer that honest conviction hesi-tates, or gives to its beliefs a timid and uncertain utter-ance.

Observe that, however man may have originated, it does not alter his position in the scale of being. It is no degra-dation to have been descended from some antecedent form of life, any more than it is an exaltation to have been formed directly from the dust of the earth. In the present state of our knowledge it is even more difficult to conceive of an origin from inorganic matter than of a development from some pre-existing form. No one can compare with atten-tion the vertebrate skeleton (of the mammals, for example) without perceiving that it is *adaptive modification* that runs throughout the whole of the ascensive orders, rather than *independent creation* of newer and higher forms. In his physical relations man is as much dependent upon external conditions as the lowest creature with which he is associated in the scheme of vitality. This scheme of vitality is one united whole, from which you cannot possibly dis-sociate any of its component parts ; and whatever be the plan that has regulated the development of this scheme in time, it must embrace man as certainly as it embraces all or any of its other members. His high position depends not so much upon the physical life which he shares in common with other beings as upon his improvable intel-lect ; and this, in any theory of his origin, can only be re-solved into a newer and higher creational endowment—the latest manifestation in that Divine plan of cosmical pro-gress which science is ever humbly and reverently endea-

vouring to reveal. We accept the fact of this new endowment; shall we reject the continuity of progress through
which it has been evolved?

Such are some of the reasonings that suggest themselves
in reviewing the question of "Man's Place in the Geological Record." In the first place, let it be treated without
bias or predilection, as a matter of natural history and
geology. In the second place, let us avail ourselves of
all the evidence that history, archæology, geology, and
palæontology can supply. And in the third place, let us,
as true geologists, be wary in assigning dates in years and
centuries, while the whole superstructure of our science is
founded on a relative and not upon an absolute chronology.
Guided by these methods, it would appear that man has
been an inhabitant of Southern and Western Europe from
a time immediately succeeding the close of the glacial
epoch, and that in these regions his antiquity dates, if not
from the very earliest, at least from the earlier of the post-
tertiary formations. How long ago this may have been in
years and centuries, there is no condescension on the part
of legitimate geology; but clearly it is far, very far, beyond
the limits of the ordinarily received chronology of the
human race. But ancient as this may be, the implement-
bearing gravels, the cave-earths, the peat-mosses, shell-
mounds, and lake-dwellings of Europe cannot be taken as a
measure of antiquity for Asia, from which, as everything
tends to show, the first races of Europe were derived by the
ordinary means of natural dispersion and selection. And
even were the first appearance of the white or Caucasian
race geologically determined in Asia, the first appearance of
the coloured varieties (Mongol, Negro, Malay, &c.), each in
its own proper headquarters, would still remain a problem
of antecedent date, requiring similar methods of research,

and similar processes of solution.   In this way, and on the fair presumption of the coloured and inferior being the older varieties, the antiquity of man as a species mounts still higher and higher, and the course of discovery may yet compel us—nay, will almost to a certainty compel us—to assign to him an origin coeval with the very dawn of what we are in the habit of regarding as the Quaternary epoch, if not, indeed, with the close of the Tertiary period, and just when the more gigantic fauna of that era were passing away from the warmer zones of Asia, Africa, and America.

ORDER AND SUCCESSION OF LIFE.

HOWEVER interesting it may be to trace the material changes
to which the crust of the earth has been subjected, this interest falls infinitely short of that excited by the study of
the life-forms by which its surface has been successively
peopled. In the one case we know something of the forces
by which the changes are produced, and the modes in
which these forces operate; in the other we perceive only
the external conditions under which plants and animals
exist, but we know nothing of the origin of Life, and as yet
very little of the causes concerned in the numerous variations and aspects it has assumed. In the one case we can,
to a certain extent, mould and modify the operating forces;
in the other, vital action is altogether beyond our production, and we can modify its variations only in the slightest
and most temporary degree. In the one case we have
masses that are operated upon from without; in the other,
forms that are actuated by impulses from within, and, in

the higher animals at least, by the subtler promptings of
the intellect and reason.   In the one case we deal with
forms and forces that are extraneous ; in the other, with
those to which we ourselves belong, and hence the higher
the interest excited, and the deeper the mysteries involved.
What is life ?   Whence its origin ?   And how has it manifested itself during the long ages throughout which geology
has traced its presence in the rock-formations of the globe ?
These are questions of the highest interest to science, and
how feeble soever the indication towards a solution which
human knowledge can offer, every tracing is of value so
long as it is founded upon fact, and sketched by an honest
hand.   Such is the aim of the present Sketch—a deduction
from the statements in the preceding chapters, a digest of
the discoveries of palæontology, an indication, if not of the
nature and relations of life, at least of its order and succession as warranted by the truths of Geology.

Concerning the origin of life, Geology ventures no opinion.   It may have started into being at the immediate
fiat of the Creator, or it may have arisen through secondary
causation, acting in obedience to Creative Law.   It may
have sprung from a single primordial germ, or it may have
spread from several primordial sources.   In its essence it
may be a thing *per se*, or it may be merely a manifestation
arising from the interactions of the subtler physical forces.
On such points Geology offers no opinion.   It deals with
life as it finds it, and dates its commencement with the
earliest traces yet discovered in the stratified formations.
At one time this limit was found in the lower Silurians,
more lately in the Cambrians, and now, as we have seen in
Sketch No. 5, discovery has carried it back to the Laurentian system—a suite of strata of much higher antiquity.
Whether these Laurentian rocks are the oldest or earliest
in which traces of life can be detected, Geology does not

aver; but as the forms of vitality become fewer and lower in kind with each successive remove in time, and as the Laurentian forms are both very lowly in organisation and scanty in number, inquirers are constrained to believe that they are nearly approaching, if indeed they have not already reached, the first beginnings of life on our planet. Be this as it may, the earliest known forms of vitality occur in the Laurentian system, and it seems something more than a mere coincidence, that these forms should belong at the same time to the very lowest orders that are known to Zoology. Such is all we know of the commencement of life on our globe; such is the ultimate limit to which geological science has yet been enabled to push her investigations. But while this is incontrovertible as matter of fact, we may believe—and all analogy seems to favour the belief—that life was contemporaneous with the laying down of the first-formed sediments. The external conditions (light, heat, cold, rains, rivers, and seas) that favour the one set of operations are usually those that accompany the other; and thus, wherever sedimentary strata occur, there also may we expect to find traces of vegetable or animal organisation. If the Laurentian be the earliest formed strata, we have already reached the goal; should others have existed before them, we can merely regard them as the provisional commencement of that long line of vital development which Geology is still labouring to reveal.

But though ignorant of the origin and commencement of life, we know something of its nature and functions. We perceive that minerals increase by the *accretion* or external addition of similar matter, but vegetables and animals grow by the internal *assimilation* of substances which they absorb and convert each into its own proper tissues. Once formed, and the course of the mineral is completed; once matured, and the plant or animal gives birth to similar plants and animals, and the course of reproduction may endure for ages. Wherever heat, light, and moisture are present, there life occurs, fitted partly for the air, partly for the land, partly for the waters, and partly also for a parasitic existence, on and within the tissues of other plants and animals. Unless under the extremes of heat and cold, life is everywhere present, restricted, no doubt, to a thin film of the globe measuring *vertically*, but spreading *horizontally* over every belt of latitude, and enjoying, each order according to its grade of organisation, the realisations of growth and reproduction. We can imagine a material world devoid of all manifestations of life, and such our planet may have been during ages of which we have no geological indication ; but, constituted as it now is, its harmonies would be incomplete without the presence both of vegetable and animal existences. Not only is the presence of the one necessary to the life of the other, but both are indispensable to the consumption and reproduction of those substances by which the structure and individuality of our globe is maintained. The crust of our earth is a thing of vegetable and animal as well as of mineral growth, and we may be assured, that from the beginning it was contemplated that each should perform its part in the harmonious maintenance of the whole. The mineral building up its chemically composite structure, the plant disintegrating and living upon these elements, the herbivorous animal feeding upon the plant, the carnivorous animal upon the herbivorous, the animal breathing the oxygen of the atmosphere and exhaling carbonic acid, and the vegetable imbibing carbonic acid and discharging in turn the oxygen, are but so many stages in a cosmical succession as harmonious in its adjustments as it seems interminable in its duration. Mineral, vegetable, and animal, are evident co-adaptations of the same great plan. We may never know how they originate, or why they exist, but we perceive the modes in which they operate, and can determine the results of their harmonious and incessant inter-actions.

At the present day we know that the great regulators of plants and animals are heat, light, and moisture. Some are adapted to the warmer regions of the globe, some to the temperate, and others, again, to the colder latitudes. Some are fitted for life on the dry land, some for life in the waters; while others, again, are fitted for both, or even to wing their way through the atmosphere. Some affect the marsh, while others cling to the thirsty upland; some rejoice in the shallow waters of the shore, while others find their fitting habitat only in the deeper ocean. Some are restricted to specific centres of limited extent, and present little variation in character, while others enjoy a wider range, and break into numerous and often widely divergent varieties. Such are the obvious conditions and distributions of life now, and such we may be certain were the nature of its conditions and dispersion sduring all previous periods. Again, some plants are rooted independently in the soil, while others find their subsistence on other plants, or even in animals. A vast number of animals live on vegetables, while others prey on the vegetable-feeders, and are fitted alone for this mode of existence. The tooth to tear is as necessary as the tooth to grind, the foot to seize as the foot to run, the hand to climb as the hand to hold, and the limb to fly as the limb to walk or the limb to swim. Such are the arrangements by which the balance and harmonies of life are now sustained, and such we may presume were the methods by which its harmonies were secured in former ages. Plant-feeder and flesh-feeder, life and death, reproduction and decay, are necessary concomitants in the great biological scheme of creation. We perceive them operating in full force now; palæontology declares they were as universally energetic in bygone epochs.

Besides these obvious arrangements and inter-relations of life, there is also grade or degree of organisation—some orders being more simple in structure and fitted for lower functions, others being more complex and fitted for the higher offices of vitality. The lichen incrusting the rock or the sea-weed clustering the shelving reefs of the sea-shore, and devoid of stem and leaf, is a lowlier organism than the clubmoss or tree-fern; and these again are less highly organised than the true timber-tree, with its complex development of trunk, branches, leaves, flowers, and fruits. The sponge, rooted plant-like to the rock, is clearly a lower phase of animal life than the coral or star-fish; and these again are less highly organised than the crabs and shell-fishes. Animals like the star-fishes, crabs, and shell-fish, devoid of backbone and bony skeleton (the *Invertebrata*), are obviously lower in the scale of being than the fishes, reptiles, birds, and mammalia (the *Vertebrata*); and even among these vertebrates the cold-blooded water-breather is inferior to the warm-blooded air-breather—the fish on a lower stage than the reptile, the reptile than the bird, and the bird than the mammal. Still more, the marsupial or pouched mammal, bringing forth immature young and carrying them about for months, is less highly organised than the placental or true mammal, which gives birth to fully developed young; and among the true mammals themselves there are manifestly various grades of organisation—the whales are lower than the ruminants, the ruminants than the carnivora, the carnivora than the monkeys, and the monkeys than man. And whatever may have been the abundance or development of life during the geologic epochs, we may presume that such grades and distinctions have ever pervaded the whole vital scheme—high and low, low and

lower, being indispensable concomitants, then as now, according to the field to be occupied and the function to be
performed.

But while there are, have been, and will continue to be such
distinctions and inter-relations of life, geology has unfolded
a new phase, unknown to, and unthought of by, our ancestors. To the earlier botanists and zoologists the existing
aspect of life seemed complete, and only such diversities
were acknowledged as those produced by station and habitat in the various latitudes of the world. Now, however,
geology has revealed by proofs innumerable and unmistakable that myriads of forms have ceased to exist, that as we
go backward in time these forms become more simple or
less highly organised, and that a period is at last reached
when our planet was peopled only by the very lowest forms
of vegetable and animal existence. The establishment of
this ascent in time from lower to higher forms is one of the
noblest triumphs of geology—a revelation that has invested
the scheme of life with new significance, and imparted to
its study a higher and more enduring interest. How broad
the Scheme of Life when living and extinct are conjoined;
how marvellous the inter-relations that subsist between its
innumerable and varied parts; and yet how ceaselessly the
whole has ever been passing onward into newer and higher
manifestations! What has been the order of this ascent?
Has it been regular and progressive? Is the higher form
linked on to the lower by characters that are invariable and
discoverable? and if so, what seems to be the law that has
regulated this progression and development? Such are
some of the questions suggested by the consideration of
this ascent of life; and as Geology has revealed the fact, it
may be legitimately allowed to attempt an explanation.

Assuming (and as yet we have no proof to the contrary)
that only the lowest or protozoan forms of life occur in the

Laurentian strata, it is admitted that we have an ascent to
corals, annelids, and crustacea in the Cambrian, and to a
still higher and more abundant display of corals, star-fishes,
shell-fishes, annelids, and crustacea in the Silurian. Not
only are the forms more numerous and varied, but at each
successive stage newer and higher orders come into view,
and we are compelled by evidence obtained from these early
systems, wherever they have been examined, to believe that
the march of life has been steadily onward and upward.
Up to this stage the living forms are wholly invertebrate
(that is, if we except a few doubtful instances in the upper
Silurians), and the plants chiefly sea-weeds and lycopods;
but in the Old Red Sandstone ferns and coniferous fragments are found, and fishes make their appearance in considerable abundance.   In the Carboniferous system, sea-
weeds, lycopods, ferns, equisetums, reeds, coniferous trees,
and numerous intermediate forms testify to a continuous
and gradual ascent in the flora ; while to a greater exuberance of all the previous fauna is added the existence of
aquatic and terrestrial reptiles.   In the earlier Secondary
rocks birds make their appearance, or at all events have not
been detected in more ancient strata; and in the upper
Secondaries, mammalian life of the marsupial orders begins
to manifest itself in increasing abundance.   It is not till
we arrive at the Tertiary system that the higher mammals
occur; and not till the Post-tertiary or Recent period that
we have any reliable evidence of the presence of Man, or of
his industrial operations.   This progressive ascent in the
animal scale is accompanied by a similar advance in the
vegetable—the palms and coniferæ of the Secondary ages
being succeeded by the exogenous or true timber-trees of
the Tertiary, and these again by the timber-trees, fruits, and
cereals of the Current epoch.   There may be imperfections
in the geological record—districts unexamined, and fossil

forms yet undetected; but the fact of this wonderful coincidence of ascent in such large areas as Europe, North America, and parts of South America, Asia, and Australia, surely entitles the belief in a progressional development of life from lower to higher forms—from thallogen to acrogen, from acrogen to endogen, and from endogen to exogen;[*] from invertebrate to vertebrate, from cold-blooded water-breathers to warm-blooded air-breathers, and from simply sentient existences to the higher activities of intellect and reason. However complicated in detail, there can be no gainsaying of this broad outline of ascent; geologists may differ in their interpretations of the means by which the phenomena have been produced, but they cannot refuse assent to the order in which they occur. Nor is it a mere ascent in general terms from lower to higher classes, but the orders, the genera, and the species partake of the same progressional development—this progression bearing a striking analogy to the order that prevails among the existing grades of life, and in most instances also to the successive stages of individual development. The forms, vegetable and animal, of the Primary periods are on the whole lower, more embryonic in aspect, and less specialised than the forms of the Secondary, and these again less than those of the Tertiary and Current epochs.[†] The ascent runs through every ordinal group and section—branching onwards and upwards; and is, we may rest assured, however little we may perceive it, as operative now as during any of the bygone geological ages.

[*] See Schemes of Vegetable and Animal Classification in Sketch No. 7.
[†] " A progress from more generalised to more specialised structures " (we quote Professor Owen, the most cautious, perhaps, of modern anatomists), " analogous to that exemplified in the existing grades of animal life and in successive phases of individual development, is appreciable in the series of species which have succeeded one another upon our planet."

As is well known to every reader of modern science, various hypotheses have been advanced to account for this progressive development of life as revealed by geology. Proceeding upon the assumption that the Scheme of Vitality is one united whole, these hypotheses endeavour to show that all the forms of plants and animals that have appeared or will yet appear are but modifications by secondary means of forms that had previously existed. Admitting that the origin of life is beyond the grasp of science, they neverthe-less seek to explain how the lowly forms of the Primary periods were developed into the higher orders of the Secondary ages, and these again into the still higher orders of the Tertiary and Current epochs. Perceiving that plants and animals do undergo modifications and variations accord-ing to climate, food, and geographical position, some would ascribe the whole development and ascent which Geology has revealed to the mere force of external conditions, oper-ating through untold ages. Others, regarding this view as inadequate, would seek to combine with the force of exter-nal conditions the use and disuse of organs, by which any organ habitually disused may become feeble, or altogether disappear, while another organ by more varied use may be increased or even transformed into an organ of altogether different form and fitted for an altogether different purpose. By these means, the plant originally formed for the waters may become fitted for the marsh, and the plant of the marsh adapted for growth and reproduction on dry land. And by the same process the limb originally destined to swim may be transformed into one fitted for walking, and the limb for walking on land converted into one adapted for flight in the atmosphere. Such changes as these, they contend, must take place by almost imperceptible stages, and must require long ages for their elaboration—the newer qualities being transmitted from generation to generation, and accelerated during the embryonic state of each successive generation.* Others, again, combine with these views the doctrine of natural selection, by which, during the incessant changes of geographical conditions, those varieties of plants and animals best fitted for the new conditions will survive and multiply, while those less adapted will gradually die out and disappear. By such a process, these theorists contend, such variations may go on till they amount to specific distinctions, and species themselves be converted into other and higher genera. Perceiving that the ascent of life in time—that is, the ascent of life as shown by Geology—coincides in a wonderful manner with the ascent from lower to higher in living plants and animals, these hypothesists admit the existence of a great creational plan, but seek to explain its development through the operation of secondary causes. Seeing that the physical phenomena of nature are brought about by the operation of secondary causes, and seeing that life is inseparably bound up with and dependent on physical conditions, they seek to apply to the one the same methods of research and reasoning which

* A little reflection will enable the non-scientific reader to perceive the force of these arguments. The eyes of animals inhabiting dark caverns gradually degenerate, and in course of generations become merely rudimentary. The feet of ruminants habitually frequenting arid plains gradually lose the digits that add to the resistance of the hoof on soft and swampy ground. Wingless birds, like the apteryx and ostrich, while they have lost the power of flight, apparently through disuse, acquire by degrees greater size and strength of limbs. Certain structures which are transitory and rudimental in existing species, are persistent and developed in extinct. Thus, the heterocercal or unequally-lobed tail, universal in palæozoic fishes, is still found in the embryo state of existing fishes, which have chiefly homocercal or equally-lobed tails. The tapering caudal vertebræ which appear in the embryo or chick of modern birds, was persistent and characteristic in mesozoic species like the *Archæopteryx*. These and many similar facts well known to anatomists give foundation and reliability to these theories, and evidently point the way to the Law which has regulated and still continues to regulate the development of organic existences.

they apply to the other, and, philosophically speaking, there is no other mode of approaching the question. We must deal with life as we find it, and however much science may fail in its demonstration, it is bound at least to make the attempt. In the present state of human knowledge, reason may be unable to grasp all the subtle and multifarious conditions that regulate the development of vitality on our globe; under a deeper and broader insight into nature's operations, the truth may begin to dawn upon us, and all the sooner and clearer the sooner and more earnestly we commence the investigation.

Admitting the philosophy of the methods, we may still be permitted to inquire how far any or all of these hypotheses are adequate to the solution of the problem? In the first place, it is obvious that external conditions operate powerfully in the distribution of plants and animals, and it is also admitted that they modify, in the long-run, the species subjected to their operation; but it seems incredible that they could transform one species into another species, or one order into another order, without the aid of other and more intimate physiological causes. Even combining the force of external conditions with the use and disuse of organs, and with the impulse of hereditary tendencies in the embryonic or fœtal stage of each successive generation, the combination seems inadequate without some other factor to keep the successive development in conformity with the known plan of vitality. Should we add the operation of natural selection, which is admittedly a powerful and ever-active agent in the modification of species, still the questions remain—how plants and animals should exist in their present ordinal arrangements, and how their appearance in time should coincide, in the main, with these ordinal arrangements? There are clearly some other factors over and above all those which have yet been brought forward, to account for the plan of vital development; some physiological law for the general order of Life analogous to that which governs the growth and development of the individual.   The better, then, that we understand the physiology of individual growth and varietal divergence—the fuller our knowledge of the palæontological order and ascent of vitality,—the sooner will we be enabled to arrive at some indication of this Law of progressional development.   If the teachings of geology are not a delusion, the fact of vital progression on this globe is as certain as the succession of its stratified formations.   Science has already done much to arrange and elucidate the one; may we not hope that under the light of increasing knowledge and more philosophical methods, human reason shall by-and-by attain to a satisfactory explanation of the other?   It is admitted that in course of time, and under new conditions, plants and animals do break into new varieties.   Variation is thus merely a matter of time and continuance of condition.   And if this be admitted, we have only to discover in the respective species those physiological peculiarities which give direction and character to the new variations.   Mysterious as the ordainings of life may seem, the problem is manifestly bound up with the operating forces of the universe, and as such is hopefully within the reach of science, as it is certainly within its legitimate domain.   All that we know of the growth, reproduction, and decay of vitality are the results of physical causation, which can be investigated and determined; shall we cease to believe that its development in time is similarly produced and as capable of demonstration?

The outcry that has been raised in certain quarters against these hypotheses of vital development, is utterly senseless and unworthy.   Investigators perceive that certain plans pervade the vegetable and animal kingdoms, and that the whole is inseparably associated in one vital

U

scheme. They perceive that life is governed in its distribution and existence by the operating forces of the universe; they learn from Geology that it has taken a certain
order of ascent in time, from lower to higher forms; and
as students of nature, they endeavour to account for its
progression by appealing to the forces by which it is manifestly affected. Philosophically there is no other course
left to them. They must deal with life as they deal with
the other phenomena of the universe; and human reason
is never more religiously occupied than when earnestly
striving to comprehend and account for the designs and
methods of the Creator. If there has been any irreverence
in dealing with this problem, that irreverence must rest
with those who would circumscribe the range of reason,
and seek, by unworthy clamour, to deter the human intellect from rising to some conception, however faint, of the
laws by which the Creator has chosen to develop the
phenomena of His marvellous universe. The authors of
these hypotheses may be right, or they may be wrong, in
their views; they may ascribe too much or too little to
certain agencies: but so long as they honestly endeavour
to arrive at the truth, their opinions ought to be gratefully
received and treated with candour. As students of science,
they abide by scientific methods; and unless Life is to be
altogether removed from the category of Natural Science,
they have no alternative but to treat it in all its bearings—
its rise, progress, and inter-relations—as they would deal
with any other problem that comes within the scope of
their investigations. Higher than mere material phenomena, more subtle in its relations than any physical
agency, more marvellous in its growth and reproduction
than any other department of nature, its comprehension, no
doubt, taxes the severest efforts of the human understanding; but still, as portion of the universe, it partakes of all

its progress, and must be amenable to all its laws. In this view, Life is simply one of the great features of the universe, carried along with it through all its mutations, and governed by the same great law of creational progression.

And this great law of creational progression is alike operative in the material, vital, intellectual, and moral phenomena of the universe. Nothing stands still. That which has been will never occur again ; that which appears now will assume a different aspect in the future. All the former distributions of sea and land, with their various surfaces, climates, and productions, have disappeared ; and the existing distribution is as incessantly passing into newer forms and aspects. All the phases of life which Geology has revealed differ with each successive formation ; higher succeeds lower at each advancing stage ; and the present, we may rest assured, will be followed by a similar progressional advancement. Man, too, in all his interrelations, is subject to the same all-pervading law. Physically, the lower variety has preceded the higher, and the highest variety of the present day stands on a lower platform than that which is destined to succeed it. Race after race has risen from barbarism to higher and higher stages of intellectualism and civilisation ; and in the moral world clearer and purer views gain, age by age, a wider recognition and more general fulfilment. Nothing stands still ; truth alone is eternal ; and as the whole world, physical, vital, intellectual, and moral, must partake of this progress, truth itself will illuminate a broader field, and lead to nobler and more godlike activities.*

* "In the lapse of ages, hypothetically invoked for the mutation of specific distinctions," says Professor Owen, " I would remark that Man is not likely to preserve his longer than contemporary species theirs. Seeing the greater variety of influences to which he is subject, the present characters of the human kind are likely to be sooner changed than those of lower existing species. And with such change of specific

Such seems to have been and to be the destined order and succession of vitality. As in existing plants and animals we perceive a vast variety of grades from lower to higher forms, so throughout the development of Life in time there has been a coincident ascent from the lower to the higher orders. During the primary periods the lands and waters were peopled only by the lowest forms; but as time rolled on, higher and higher orders gradually made their appearance, each successive rock-system bearing testimony to the introduction of newer and more highly organised existences. Imperfect as the geological record admittedly is, and limited as may be the portions of the earth's crust yet examined, there is such a coincidence in the fossil life of all the surveyed tracts, that we may regard the order of ascent as an established fact, subject only to minor modifications in the details. The differences that have arisen among geologists relate not to the facts of the ascent, but to the mode or modes in which the development has been brought about. As we know more of extinct life and more of the physiology of existing life, these differences will in a great measure disappear, and conflicting hypotheses give way to a uniform and satisfactory theory. In the mean time every earnest endeavour is entitled to our regard, and however startling its views, or how little soever it may seem to clear the way to sounder conclusions, it ought to be gratefully received as a contribution towards the solution of the highest and most interesting problems, perhaps, that the progress of discovery has submitted to the consideration of modern philosophy.

character, especially if it should be in the ascensive direction, there might be associated powers of penetrating the problems of zoology, so far transcending those of our present condition as to be equivalent to a different and higher phase of intellectual action, resulting in what might be termed another species of zoological science."—*Preface to Comparative Anatomy and Physiology of Vertebrates.*

WHAT OF THE FUTURE?

POSSIBILITY OF INDICATING THE FUTURE OF OUR PLANET—OPINION
OF DR HUTTON—EVERYTHING IN NATURE, PHYSICAL AND VITAL,
PASSING ON TO NEWER FORMS AND CONDITIONS—NEW DISTRI-
BUTIONS OF SEA AND LAND—NEW CLIMATES AND PHYSICAL
SURROUNDINGS—NEW ARRANGEMENTS OF PLANTS AND ANIMALS
—NEW DEVELOPMENTS, OR HIGHER AND HIGHER LIFE-FORMS—
MAN SUBJECT TO THE SAME LAW OF PROGRESSION—INFLUENCE
OF MAN ON THE FUTURE—SLOW AND GRADUAL RATE OF NATURE'S
OPERATIONS—EXALTED CONCEPTIONS OF THE UNIVERSE INSPIRED
BY THE BELIEF IN A LAW OF INCESSANT DEVELOPMENT AND
PROGRESS.

" MAN," says Dr Hutton in his celebrated ' Theory,' " first
sees things upon the surface of the earth no otherwise than
the brute, who is made to act according to the mere impulse of his sense and reason, without inquiring into what
had been the former state of things, or what will be the
future. But man does not continue in that state of ignorance or insensibility to truth ; and there are few of those
who have the opportunity of enlightening their minds with
intellectual knowledge, that do not wish at some time or
another to be informed of what concerns the whole, and to
look into the transactions of time past, as well as to form
some judgment with regard to future events. It is only
from the examination of the present state of things that
judgments may be formed, in just reasoning, concerning
what had been transacted in a former period of time ; and

it is only by seeing what had been the regular course of
things that any knowledge can be formed of what is after-
wards to happen; but, having observed with accuracy the
matter of fact, and having thus reasoned as we ought, without supposition or misinformation, the result will be no
more precarious than any other subject of human understanding." No more precarious than any other subject,
and a great deal more certain, indeed, than most of the
topics with which the human understanding is apt to busy
itself! Here is a world, Physical Geography informs us,
having certain ordainings at *present;* here is a world, Geology informs us, which has had a strange and varied history
in the *past;* and combining our knowledge of past and
present, with faith in the uniformity of nature's operations,
we are surely entitled to speculate with some degree of certainty as to the fate that awaits it in the *future.* Such spec-
ulation forms the subject of the present Sketch, and guided
by the spirit of the Huttonian philosophy as above expressed, we do not despair of arriving at something like an
intelligible indication. This indication may not exactly
carry us forward to positive appearances, but it will show
us, at least, what cannot continue, and thus the better pre-
pare us for the perception and admission of the changes
that must follow. No doubt, the changes in the natural
world are so multiform and complex, and their producing
causes act and react so unequally upon each other, that
man, limited in his faculties and imperfect in his knowledge, can never hope to forecast either their exact order
or amount; still, by adhering to right methods, he can
sketch an outline of the future, just as he has been enabled
to trace the vestiges of the past, and this outline, shadowy
as it may be, is something at least for Geology to boast of.
It is true that it will add no new fact to our knowledge,
for facts are things that have been accomplished; but its

tracing gratifies the intellect, and the belief in its certainty exalts our conceptions of the course and continuity of creation.

And, *first*, we may safely assert that the present distribution of sea and land, with all its diversity of continent and island, will not be the prevailing arrangement in future ages; and that the more remote the period, the greater in all likelihood the difference. All that Geology teaches of the past, shows that sea and land have repeatedly changed places; all that Physical Geography tells of the present, declares that similar changes are incessantly in progress. Every wind that blows, every frost that freezes, every shower that falls, river that runs, and wave that strikes, is wasting and wearing down the framework of the existing continents; and the eroded material, borne down to lakes and estuaries and seas, is gradually displacing so much of the water and creating newer lands. This waste is so apparent on every cliff, and on every ravine and river-glen, that its truth requires no further enforcement; and the same holds good of every shore against which the waves dash, or the tidal currents scour. But while loss goes forward in one region, gain takes place in another; and thus most of our river plains are but the sites of silted-up lakes, just as all our deltas, occupying millions of square miles, are recent and still progressing acquirements from the sea. Nor is it alone to forces from without that the surface of our earth is subjected. The forces from within—the volcano, earthquake, and crust-motions, described in Sketch No. 3—are equally active and incessant; here piling up new hills, there throwing up the sea-bed into dry land, and here, again, submerging terrestrial surfaces beneath the waters of the ocean. If, then, such changes are unmistakably taking place at the present day, and if by a parity of reasoning we can show that kindred changes took place in times past, we are surely entitled to rely on the uniformity of nature's operation, and to believe that similar changes will continue to be effected. The conclusion is irresistible, and thus must be admitted the truth of our first proposition, that the future distributions of sea and land must differ from the present, and that as time rolls on the divergence will become greater and greater, till all the existing continents disappear, and new ones arise, with other contours, other surfaces, and other climates, but all fluctuating and progressive as those that went before. The whole history of the past, as interpreted by Geology, has been one incessant round of terraqueous change ; the forces of nature are still as active and unabated in operation ; and the inevitable results must be a round of terraqueous changes in the future, as incessant in their recurrence and as extensive in their range.

In the *second* place, it is equally clear that if the lands of the current era are gradually disappearing, and newer ones as gradually in course of formation, the latter must occupy different positions on the earth's surface, and as a consequence, must enjoy different climates and different geographical surroundings. They may be more continental or more insular, more tropical, more temperate, or more arctic; but however this may be, it is quite evident they must differ in these respects from the existing. In ignorance of the law which regulates the great crust-motions of the globe—those slow upheavals and submergences of extensive tracts—we can scarcely indicate the position of the lands that will immediately succeed the existing; though, judging from the directions of most of the great delta-forming rivers, it would seem that they will lie more within the warmer zones of the globe. The great rivers of the New

World—the Mississippi, Orinocco, Amazon, and La Plata —are all projecting their vast deltas within these warmer zones; so also are the larger rivers of Europe and Africa, for though the Nile has a northerly trend, its delta is still within these warmer limits; and so also are the more important delta-formers of Asia—the Tigris and Euphrates, the Indus, Ganges, Irawaddy, Menam, Kong, Yang-tse-Kiang, and Hoang-ho. One has only to cast his eye over a map of the " river-systems," to perceive the truth of this assertion; for though Siberia and Arctic America are also possessed of large rivers, their ice-bound character and other conditions render their land-forming powers comparatively unimportant. To this southerly projection of the great rivers must also be added the fact, that vulcanic agency is much more active in its accumulations within tropical and sub-tropical than within arctic or antarctic zones. Appeal to any volcanic map of the world, and see how blank and quiescent are the great tracts of North America, Northern Europe, and Northern Asia, as compared with the warmer and more southerly zones, and then judge how much this may have to do with the formation and disposition of the future continents. In all the formations of newer lands vulcanicity has played an essential part; can we fail to infer that the vast cincture of volcanic action that surrounds the Pacific, as well as the numerous centres that stud its basin, are intimately connected with the elaboration of future land-masses within its area? It is true that the gradual elevation noted by voyagers among the Arctic Islands, Northern Greenland, Spitzbergen, Scandinavia, and Siberia, would seem to point to a more continuous massing of the land in that direction; but though this were to be the case, it would not interfere with the likelihood of our previous surmise, that the dry lands more immediately succeeding the existing will lie within more southerly and warmer latitudes. But the mere lying within more southerly latitudes does not altogether determine the nature of a climate or the geniality of external conditions. Insular or continental arrangement of the land-masses, altitude above the sea-level, the direction of mountain-ranges and valleys, the amount of rainfall, and the set of ocean-currents from colder or warmer regions, all influence less or more the climate of a country; and these, as affecting the future lands of our globe, we have no means whatever even of guessing at. All that we know for certain is, that the existing continents are gradually passing into newer forms and dispositions, and that these dispositions must of necessity be accompanied by other climates and by other vegetable and animal appointments. All that we can offer as a legitimate geological inference is, that the land-masses more immediately succeeding the existing will be more sub-tropical and tropical in their position, and, *cœteris paribus*, more congenial in their physical surroundings.

In the *third* place, if the present continents and islands are to be superseded by others possessed of different external conditions, it follows that these newer lands must be characterised by different distributions of plants and animals. At present it is presumed that each plant and animal occupies the station and habitat best suited to its growth and perfection; but if these situations be interfered with, no matter how slowly, certain races must succumb to the change, while others will usurp their places. Were the future continents, for example, to be gradually evolved within lower latitudes, their flora and fauna would as gradually assume a tropical aspect—existing forms, on the other hand, disappearing stage by stage with the external conditions to which they had been originally adapted. The disposition of the existing continents is mainly longitudinal, hence they

are subjected to extreme diversity of climate, and consequent variety of vegetable and animal life; were the lands
of the future to be arranged more latitudinally, a greater
uniformity of climate would prevail, and with this uniformity a corresponding sameness in the specific distribution
of vitality. The existing land-masses bulk most largely in
the northern hemisphere; we can readily perceive how different the climatic and vital arrangements of the globe
were they mainly disposed in the southern. The trade-
winds, tides, and oceanic currents—the great modifiers of
climate—are interrupted in their normal continuity by the
longitudinal disposition of the existing continents, and
thrown into minor and complex deviations; how different
the result did the disposition of the land permit them to
revolve in simple and unbroken continuity! On the whole,
it may be safely assumed, that the greater the difference between future and existing continents in position and climate,
the greater will be the difference in their vital appointments;
and thus the future, from those physical causes alone, would
present a totally different life-picture from the present.
There might be no new creations nor developments, but
there would be extensive re-arrangements and re-assortments
of the existing—extermination of certain genera and species
and increase of others, and along with these a corresponding
redistribution of the varieties of the human family itself.
We have said "there might be no new creations nor developments," and yet it is difficult to conceive of any extensive alterations in the distribution of sea and land, without
associating with them newer phases of vegetable and animal
existence. In the history of the past, new developments of
life are so intimately associated with geographical changes,
and every newer formation so distinctively characterised by
its own peculiar and higher species, that whatever the law
by which the succession of vitality on our globe is gov-

erned, external conditions are undoubtedly one of its principal factors; and thus, with other distributions of sea and land, we may anticipate not only other distributions, but other and newer phases of their flora and fauna.

But, *fourthly*, if we are to be guided in our speculations respecting the future by our knowledge of the past, another and more important element comes to be considered, and that is, the continuous ascent from lower to higher life-forms which is traceable throughout the whole of the geological periods. If there has been such an ascent throughout the past by some process of development, or whatever it may be, the reasonable presumption is, that the same process will continue to manifest itself in the future. It is by no means asserted that investigators have anything like completed the geological record; but all that has been done, whether in Europe, Asia, America, or Australia, tends to the establishment of an ascent in the main—from the cryptogam to the phanerogam, from the invertebrate to the vertebrate; from the endogen to the exogen, from the fish to the reptile, from the reptile to the bird, and from the bird to the mammal. Even within these great sections there has been a corresponding ordinal advance; and although new discoveries occasionally compel us to modify our views as to the time when certain orders and genera made their first appearance, still no discovery has ever been made that militates in the least against the general doctrine, that the more lowly organised have regularly preceded the higher and more specialised orders. We say " regularly preceded," for though systematists are occasionally puzzled with apparent breaks in the continuity, these breaks are either the result of local obliterations, or they arise from limited and imperfect observation. Such having been the progressive evolution of life during the long and physically-varying cycles of the past, the geologist is surely entitled to presume that a similar evolution will continue to mark the onward course of creation.

It is true, it may be argued, and indeed has been argued, that the system of Life has culminated in the present epoch, and is, consequently, subject to no further development; and if it has not so culminated, that new races ought now and then to be making their appearance. Such reasoning, however, is altogether at variance with the slow and gradual evolution of events as impressively taught by Geology. During the past, well-defined genera and orders make their appearance only after the lapse of ages; and two thousand or twenty thousand years of the current epoch may be too short a period for the full development of new and higher races. All that seems necessary for our argument is, that physiology can prove *a tendency to variations* in existing genera and species; and if such a tendency can be demonstrated, no matter how slight and slow, the widest subsequent divergence, even to the extent of new families and orders, is only a question of time and continuation. This is all that Geology contends for; and surely the variations in plants and animals which are continually taking place under change of external condition, and under the influence of culture and domestication, must be sufficient to convince every mind capable of ordinary reasoning, that the susceptibility to newer developments is a quality as operative now as during any of the former epochs. It is vain to argue that no introduction of newer forms has been witnessed during the last three or four thousand years of human history. It is little more than half a century since such questions began to attract the attention of philosophers; and all that went before—erroneous notions of natural history, limited acquaintance with the geography of the world, ignorance of geology, and total absence of all reliable record—renders any arguments founded on this ground utterly idle and worthless. If vast physical changes have passed unmarked by our ancestors, what marvel need there be that the minuter phenomena of vital variations should have wholly escaped their attention? If the variations, descent, and dispersion of the human family be a matter of doubt and uncertainty to historians and ethnologists, what marvel need there be that the varieties, descent, and dispersion of the lower races should have passed unnoticed and even unsuspected? But if the introduction of new genera and species cannot be positively proven, we know that numerous forms have disappeared from certain localities, and that several (the dinornis, æpiornis, dodo, solitaire, great auk, rhytina, &c.), within a comparatively recent period, have become altogether extinct. As extinction and creation ever went side by side in the past, so the fair presumption is that extinction is attended by a similar creation in the present. The minutest scrutiny can detect no decay in the physical accompaniments of life, no decline in the powers of vitality itself, and no change whatever in any of its discoverable relations to external nature; and surely on these, as on all the grounds formerly mentioned, we are entitled to believe in a continuance of vital development, as firmly as in a continuance of the physical changes which are daily and hourly taking place around us. The one may be less perceptible than the other, but not on that account the less real; slower in their rate of progress, but not the less certain and continuous.

Such a progression being granted, the Life of the future must differ from that of the present, as that of the present differs from all that went before. Old genera and species must pass away, and newer and higher ones take their places. As the ratio of development among the different classes and orders, both of plants and animals, seems to have differed during the past—the lower being more persistent, or less variable, than the higher—so the same ratio will manifest itself in the future, the more highly organised passing by more rapid stages into newer and higher forms. Within the same limit of time the invertebrate may undergo less modification than the vertebrate, the aquatic less than the terrestrial, and the cold-blooded less than the warm-blooded; but all (by whatever process) must sooner or later pass into newer and higher forms, and man himself as certainly as the plants and animals that form the theme of his deliberations. If there exist a great Law of Progression—and all that palæontology has revealed of the past or physiology taught of the present points to such an ordaining—it would be reversing our ordinary ideas of the permanence of nature to suppose such progression had come to an end when all its accompaniments and all the media through which it manifests itself remain unimpaired and persistent. And it would be equally at variance with all philosophical notions of natural history to suppose that a law which had operated alike on all the orders of life in the past would become partial and exempting in the present. It may startle some minds to hear the same argument applied to man as to the rest of the animate creation. The reverse, unfortunately, has been too long the fashion. The interests of science, as well as the sacredness of truth, demand, however, that where nature operates alike the student of nature shall not venture to obtrude with artificial "schemes of arrangement," and "lines of demarcation." Looking at it merely as a question of natural history—and this, be it observed, is the only way in which science can approach it —the future of the human race must be subject to the same laws of variation and progression as those which have governed and still continue to govern the extinctions and evolutions of the other orders of vitality. The rate of

variation may differ in the respective orders, but from the
law of variation there can be no possible exemption. And
if this rate of variation has hitherto been more rapid in the
higher than in the lower orders, the existing varieties of
man must consequently the sooner pass into other and
higher varieties.

In the *last* place, a new and important element must
ever be taken into account in all our reasonings respecting
the future flora and fauna of the globe. During the old
geological epochs the " Law of Development," or whatever
else it may be termed, exerted itself universally, and this
without control or interference by human operations.
Now, however, and during the current epoch, man comes
in as a modifying and sub-creative agent—here removing
and extirpating, there transferring and multiplying, and
this the more signally the more settled and civilised he becomes. Observe what changes must have taken place since
civilised man first took possession of Asia and Europe, and
how much more since, under the impulse of modern progress, he has carried his efforts to America and Australia !
The natural flora of a region must make way for his cultivated plants, and the fauna for his domesticated animals.
Here he reclaims and extirpates, and this extirpation reacts
in a hundred ways on the surrounding fauna. There he
transfers and acclimatises, and in a few generations the
plants and animals of the Old World flourish and multiply
in the New, and by reciprocation those of the New World
become equally prolific in the Old. Here he destroys the
plant, and the animal deprived of its food shifts ground or
dies out ; there he introduces a new animal, and, in the
struggle for existence, some weaker and less elastic species
succumbs before the intruder. Of course, there are great
climatal limits to this transference, which the utmost inge-

nuity of man cannot pass, but there is no possible end to his extirpations and modifications; and thus in the lapse of time the natural course of vital development and distribution will be extensively interfered with by the operations of man. But extensive as may be his interference, he cannot overrule the great Law of Progression any more than he can prevent the physical operations by which the continents themselves are continually modified. The law will go irresistibly forward, carrying man in his several varieties along with it; employing, it may be, his influence as a pre-ordained portion of its operations, but still methodically ascending from higher to higher, till the future aspects of life differ as widely from those of the present, as its present aspects differ from those that preceded.

Such are the leading features of the future, which, reasoning from the order of the past and the appointments of the present, Geology is enabled to indicate—new distributions of sea and land, new climates and physical surroundings, new and higher orders of plants and animals; but all these brought about so slowly and gradually that the changes become apparent only after the lapse of ages. The present, so far as we can judge, is as fluctuating as the past, and yet the oldest scenes of human history—India, Mesopotamia, the valley of the Nile, the shores of the Levant, Greece, and Italy—remain in all their broader features much as they were three thousand, four thousand, and six thousand years ago. So tranquil, indeed, are the great *physical* progressions of nature that it requires some mental effort to perceive them; so gradual the *vital*, and through so many intermediate stages, that it demands an exercise of reason to admit their reality. But when once perceived, how enlarged our conceptions of the universe—a system of incessant fluctuation and progress in its details; a system

of stability and permanence in its general appointments !
To the ignorant mind, alluded to by Dr Hutton in our opening quotation, the earth is a mere monotonous panorama
of birth, progress, and decay—the same now as it has been
from the beginning, and as it is now so to continue unchanged and unchangeable to the end.   To the enlightened
mind, on the other hand, it becomes a scene of incessant
development and progress—multiform and variable in its
physical relations, diversified and progressive in its vital
appointments, and still at every turn assuming more wonderful and more exalted aspects.   How much more en-
nobling this conviction of our planet's incessant mutation
and progress than the old belief in its stereotyped sameness and ever-threatened decay !   How enlarged the conceptions of Creative Wisdom inspired by a knowledge of
these ever-varying and ever-advancing aspects—these endless adaptations and boundless resources !   And how much
more when we carry our views from this world of ours to
the other members of the planetary brotherhood, and believe them subject to similar laws, and characterised by
similar appointments !   To the eye of sense they are mere
balls swinging clearly and coldly in space ; to the eye of
enlightened reason they become, like their sister orb, multiform in the details of their terraqueous surfaces, variable in
their physical and vital aspects, and yet throughout all
their variability invariably conforming to a higher law of
development and progress.

INDEX.

THE END.